Mobile Robot Navigation with Intelligent Infrared Image Interpretation

William L. Fehlman II · Mark K. Hinders

Mobile Robot Navigation with Intelligent Infrared Image Interpretation

Springer

Asst.Prof. William L. Fehlman II
Dept. Mathematical Science
United States Military Academy
NY 10996 West Point
USA
william.fehlmanii@us.army.mil

Prof. Mark K. Hinders
Dept. Applied Sciences
College of William & Mary
P.O.Box 8795
VA 23187-8795 Williamsburg
USA
hinders@as.wm.edu

ISBN 978-1-84882-508-6   e-ISBN 978-1-84882-509-3
DOI 10.1007/978-1-978-1-84882-509-3
Springer Dordrecht Heidelberg London New York

British Library Cataloguing in Publication Data
A catalogue record for this book is available from the British Library

Library of Congress Control Number: 2009930031

© Springer-Verlag London Limited 2009

Apart from any fair dealing for the purposes of research or private study, or criticism or review, as permitted under the Copyright, Designs and Patents Act 1988, this publication may only be reproduced, stored or transmitted, in any form or by any means, with the prior permission in writing of the publishers, or in the case of reprographic reproduction in accordance with the terms of licenses issued by the Copyright Licensing Agency. Enquiries concerning reproduction outside those terms should be sent to the publishers.
The use of registered names, trademarks, etc., in this publication does not imply, even in the absence of a specific statement, that such names are exempt from the relevant laws and regulations and therefore free for general use.
The publisher makes no representation, express or implied, with regard to the accuracy of the information contained in this book and cannot accept any legal responsibility or liability for any errors or omissions that may be made.

Cover design: eStudioCalamar, Figueres/Berlin

Printed on acid-free paper

Springer is part of Springer Science+Business Media (www.springer.com)

Preface

This book describes in detail a physics-based adaptive Bayesian pattern classification model that uses a passive thermal infrared imaging system to automatically characterize non-heat generating objects in unstructured outdoor environments for mobile robots. In the context of this work, non-heat generating objects are defined as objects that are not a source for their own emission of thermal energy, and so exclude people, animals, vehicles, etc. The resulting classification model complements an autonomous bot's situational awareness by providing the ability to classify smaller structures commonly found in the immediate operational environment. Since GPS depends on the availability of satellites and onboard terrain maps which are often unable to include enough detail for smaller structures found in an operational environment, bots will require the ability to make decisions such as "go through the hedges" or "go around the brick wall." A thermal infrared imaging modality mounted on a small mobile bot is a favorable choice for receiving enough detailed information to automatically interpret objects at close ranges while unobtrusively traveling alongside pedestrians. The classification of indoor objects and heat generating objects in thermal scenes is a solved problem. A missing and essential piece in the literature has been research involving the automatic characterization of non-heat generating objects in outdoor environments using a thermal infrared imaging modality for mobile bots. Seeking to classify non-heat generating objects in outdoor environments using a thermal infrared imaging system is a complex problem due to the variation of radiance emitted from the objects as a result of the diurnal cycle of solar energy. The model that we describe will allow bots to "see beyond vision" to autonomously assess the physical nature of the surrounding structures for making decisions without the need for an interpretation by humans.

The approach described here is an application of Bayesian statistical pattern classification where learning involves labeled classes of data (supervised classification), assumes no formal structure regarding the density of the data in the classes (nonparametric density estimation), and makes direct use of prior knowledge regarding an object class's existence in a bot's immediate area of operation when making decisions regarding class assignments for unknown objects. We have used a mobile bot to systematically capture thermal infrared imagery for two categories of non-heat generating objects (extended and compact) in several different geographic locations. The extended objects consist of objects that extend laterally beyond the thermal camera's lateral field of view, such as brick walls, hedges, picket fences, and wood walls. The compact objects consist of objects that are completely within the thermal camera's lateral field of view, such as steel poles and trees. We used these large representative data sets to explore the behavior of thermal-physical features generated from the signals emitted by the classes of objects and design our Adaptive Bayesian Classification Model. We demonstrate that our novel classification model not only displays exceptional performance in characterizing non-heat generating outdoor objects in thermal scenes but it also outperforms the traditional KNN and Parzen classifiers.

We are grateful for the support received by several institutions in carrying out the work presented in this book. We are thankful for a grant of computer time from the DoD High Performance Computing Modernization Program at the Army Research Laboratory Major Shared Resource Center, using computational facilities at The College of William & Mary which were enabled by grants from Sun Microsystems, the National Science Foundation, and Virginia's Commonwealth Technology Research Fund, and the General Omar N. Bradley Research Fellowship in Mathematics provided by the Omar N. Bradley Foundation. Particularly, we thank Juan Chaves from the Ohio Supercomputer Center (OSC), and Stephen Landowne, from the United States Military Academy, for making it possible to use the DoD high performance computing system. We thank Chris Bording in assisting us with porting computer code to the computational facilities at The College of William & Mary. Additionally, we thank the U.S. Army Research Laboratory's Weapons and Materials Research Directorate (WMRD) at Aberdeen Proving Ground, Maryland, for their collaborations during this work.

We are also thankful to our colleagues and friends who have provided assistance and made comments on various parts of the manuscript. In particular, we would like to thank Darryl Ahner, Cara Campbell, Hilary DeRemigio, Danielle Dumond, Valerie Fehlman, Tina Hartley, Alex Heidenberg, Michael Jaye, Scott Nestler, Zia-ur Rahman, Leah Shaw, Jonathan Stevens, Eugene Tracy, and Deonna Woolard.

The views expressed in this book are those of the authors and do not represent the views of the U.S. Government.

List of Symbols

The following is a list of symbols used in this book.

C	specific heat ($kJ \cdot kg^{-1} \cdot {}^\circ C^{-1}$)
${}^\circ C$	degrees Celsius
$comp_{\underline{e}_{1j}} \tilde{\underline{f}}_{ij}$	component (or scalar projection) of the pattern $\tilde{\underline{f}}_{ij}$ onto the first principal eigenvector $\underline{e}_{1j}$
$\underline{D}_n$	feature vector generated from an unknown target's signal received by sensor n
D_{ij}	normal distance between a pattern i and first principal eigenvector $\underline{e}_{1j}$
$\underline{e}_{1j}$	first principal eigenvector of object class O_j
${}^\circ F$	degrees Fahrenheit
H	Shannon's entropy
h	parameter for Parzen Classifier
h_c	free convection coefficient

List of Symbols

K	parameter for *K*-Nearest-Neighbor Classifier
K_x, K_y, K_z	in-plane and transverse thermal conductivity of the object $\left(W \cdot m^{-1} \cdot {}^\circ C^{-1}\right)$
L_o, L_r	thermal radiance emitted by an object and reference emitter and detected by a thermal infrared camera $(W \cdot m^{-2} \cdot sr^{-1})$
L_b	irradiance energy on a target from the surrounding background environment $(W \cdot m^{-2} \cdot sr^{-1})$
$\mathcal{L}\left(\underline{\tilde{f}} \mid O_j, d_j\right)$	likelihood function weighted by the distance function $d_j\left(\underline{\tilde{f}}, \underline{e}_{1j}\right)$
N_j	total feature vectors from object class O_j's data set
O_j	object class with index j
$P(O_j)$	unconditional probability
$P(O_j \mid \underline{D}_n)$	conditional probability
$P(\underline{D}_n, O_j)$	joint probability
$p(x)$	probability density function
Q	heat flux $(W \cdot m^{-2})$
T, T_o, T_b, T_a	temperature; object surface temperature; background surface temperature; ambient temperature
t, t_r	time; relaxation time
V	volume of a hypersphere

Greek Letters

ε, ε_o, ε_r	emissivity; object emissivity; reference emitter emissivity
μ_n	central moments (nth moment about the mean)
μm	micro-meters (also called microns)
ρ	density $\left(kg\ m^{-3}\right)$
σ	Stephan-Boltzmann coefficient
τ	atmospheric transmission coefficient

Feature Labels

$Co1$	contrast, first-order statistic (macro feature)
$Co2$	contrast, second-order statistic (macro feature)
$Cr2$	correlation, second-order statistic (macro feature)
$En1$	entropy, first-order statistic (macro feature)
$En2$	entropy, second-order statistic (macro feature)
Eo	emissivity (micro feature)
$Er2$	Energy, second-order statistic (macro feature)
$Ho2$	homogeneity, second-order statistic (macro feature)
Lb	background irradiance (micro feature)
Lo	object surface radiance (micro feature)
Lob	Lo/Lb (micro feature)
Lor	Lo/Lr (micro feature)

Lr	reference emitter radiance (micro feature)
$Mo1$	object scene radiance, first-order statistic (macro feature)
$Mob1$	Mo1/Lb, first-order statistic (macro feature)
$Mor1$	Mo1/Lr, first-order statistic (macro feature)
$So1$	smoothness, first-order statistic (macro feature)
$T1$	ambient temperature rate of change (meteorological feature)
Ta	ambient temperature (meteorological feature)
$To1$	third moment, first-order statistic (macro feature)
$Uo1$	uniformity, first-order statistic (macro feature)

Superscripts

$\overline{O}$ verbar	sample mean (or average)

Subscripts

$\underline{U}$ nderbar	vector

List of Figures

Figure		Page
1.1	Unstructured environments as potential areas of operation for autonomous robots.	2
1.2	Visible and thermal images of a wooden fence. (a) visible image of the fence during the day, (b) visible image captured at 2030 hrs on 7 September 2007 with light source illuminating on the fence, (c) thermal image of the fence captured at the same time as the visible image in (b) and at an ambient temperature of $71.9°F$.	4
1.3	Mobile robotic 3D sonar scanning system, *rWilliam* (on right) and thermal imaging system, *rMary* (on left).	5
1.4	Thermal scene consisting of heat and non-heat generating objects. Heat generating objects include the human walking on the sidewalk and squirrel running from behind the tree. Non-heat generating objects include the trees and steel pole used by the street light.	6
1.5	Geometric measurements generated from thermal images of heat generating objects for classification. (a) measurements generated to classify people [2]. (b) measurements generated to classify vehicles [3].	7
1.6	Roomba vacuum cleaning robot.	8
1.7	Robotic lawn mower.	9

List of Figures

Figure		Page
1.8	Autonomous unmanned ground vehicle platforms designed to support various military and commercial applications. (a) military reconnaissance application (Courtesy of globalsecurity.org), (b) Battlefield Extraction Assist Robot for ambulatory applications (Courtesy of Vecna Technologies, Inc.), (c) remote monitoring and surveillance applications (Courtesy of MobileRobots, Inc.).	10
1.9	Infrared range sensor with detection range from 1 to 5.5 m (Courtesy of Acroname, Inc.)	12
1.10	Spectral radiance of a blackbody. Long-wave infrared band (7–14 microns) is denoted by the blue shaded region.	17
1.11	Pattern classification model design cycle.	20
1.12	Intelligence algorithm with pattern classification model.	21
2.1	Robotic thermal imaging system hardware: (a) robot platform front view, (b) robot platform rear view, (c) Raytheon thermal imaging video camera, (d) VideoAdvantage USB video capture device, (e) Samsung tablet PC w/ Powerbank.	26
2.2	Thermal image prior to preprocessing.	28
2.3	Control IR Manager main menu.	29
2.4	Control IR Manager video settings.	30
2.5	Control IR Manager advanced video settings.	31
2.6	Thermal image with preprocessing on temporal/spatial signal degradations and dead pixels. AGC is enabled.	32
2.7	AC coupling. (a) Scene with different temperature regions, (b) Gray-level shades of regions in thermal image.	33–34
2.8	Enabled AGC experiment with cardboard tubes (left tube at constant temperature of ~ 86.5 deg F and right tube heated to 110.8 deg F and allowed to cool to 65.8 deg F). (a) Image of tubes with right (heated) tube at 110.8 deg F, (b) Image of tubes with right (heated) tube at 65.8 deg F, (c) Variations of gray-levels of constant and heated tubes as a function of temperature.	36

List of Figures   xiii

Figure		Page
2.9	Disabled AGC experiment with cardboard tubes (left tube at constant temperature of ~ 86.5 deg F and right tube heated to 110.4 deg F and allowed to cool to 65.8 deg F). (a) Image of tubes with right (heated) tube at 110.4 deg F, (b) Image of tubes with right (heated) tube at 65.8 deg F, (c) Variations of gray-levels of constant and heated tubes as a function of temperature.	37
2.10	Thermal image with preprocessing on temporal/spatial signal degradations and dead pixels. AGC is disabled.	38
2.11	Thermal image of segment of brick wall: (a) without high pass filter, (b) with high pass filter.	39
2.12	(a) Robotic thermal imaging system capturing an image of a wood fence. (b) Thermal image of the wood fence displayed with *VideoAdvantage* software.	40
2.13	Visible and thermal images of extended objects from the training data set. (a) brick wall, (b) hedges, (c) wood picket fence, and (d) wood wall.	43
2.14	Visible and thermal images of compact objects from the training data set. Steel poles: (a) brown painted surface, (b) green painted surface, (c) octagon shape w/ aged brown painted surface. Tree: (d) basswood tree, (e) birch tree, (f) cedar tree.	44
2.15	Ambient temperature distributions for training, test, and blind data collected from 15 March to 5 November 2007.	45
3.1	Thermal Image Representation: (a) sources of radiance emitted from fence segment and received by the camera, (b) thermal image of fence segment, (c) data array of gray-level intensities from segment of thermal image.	52
3.2	Aluminum plate with low emissivity. (a) visible image of aluminum plate. (b) thermal image of aluminum plate.	56
3.3	Glass plate with high emissivity and opaque to IR radiation. (a) visible image of glass plate in front of pine tree log. (b) thermal image of glass plate in front of log. (c) thermal image of log without glass plate in front.	57
3.4	Variation of emissivity with viewing angle for a number of (a) nonmetallic and (b) metallic materials. [46]	58

List of Figures

Figure		Page
3.5	Variation of emissivity with object shape and surface temperature.	59
3.6	Directional variation of emissivity for a pine tree log outdoors. (a) experimental setup, (b) pine tree log with brick wall irradiance, (c) pine tree log with dry wall irradiance. (d) gray-level comparisons of brick wall vs. dry wall.	61
3.7	Halo effect resulting from a (a) "hot" target and "cold" foreground and (b) "cold" target and "hot" foreground.	62
3.8	(a) Thermal radiance received by the thermal imaging camera. (b) Thermal image of cedar tree captured at 0545 hrs on 17 March 2006.	64
3.9	Visible and thermal images of objects captured on 10 February 2007 to evaluate the emissivity feature. (a) steel pole, (b) birch tree log, (c) concrete cylinder, (d) hedges, and (e) wood wall.	66
3.10	Gray-level Co-occurrence Matrix. (a) spatial relationship of neighboring pixels, (b) gray-level array of a thermal image, (c)–(f) GLCMs with distance $D = 1$ and directions 0, 45, 90, and 135 degrees, respectively.	75
3.11	Visible and thermal images of extended objects used for pixel distance analysis and selection. (a) brick wall, (b) hedges, (c) picket fence, and (d) wood wall.	79
3.12	Extended objects pixel distance analysis. Pixel Distance vs. (a) Contrast2, (b) Correlation, (c) Energy, (d) Homogeneity, (e) Entropy2.	80
3.13	Extended objects absolute sum of the differences for Energy and Entropy2 features as a function of pixel distance (D).	81
3.14	Visible and thermal images of compact objects used for pixel distance analysis and selection. (a) brown steel pole, (b) green steel pole, (c) octagon steel pole, (d) basswood tree (e) birch tree, (f) cedar tree.	82
3.15	Compact objects pixel distance analysis. Pixel Distance vs. (a) Contrast2, (b) Correlation, (c) Energy, (d) Homogeneity, (e) Entropy2.	83

Figure		Page
3.16	Compact objects absolute sum of the differences for Energy and Entropy2 features as a function of pixel distance (D).	84
3.17	Visible and thermal images of objects used to evaluate thermal features. Extended objects: (a) brick wall, (b) hedges, (c) wood wall. Compact objects: (d) concrete cylinder, (e) steel pole, (f) pine tree log.	85
3.18	Visible and thermal images of objects used to demonstrate curvature algorithm. Segmented regions in thermal images are used to compute the average radiances used in the curvature algorithm. (a) tree, (b) square metal pole, (c) brick wall.	90
4.1	Scatter plot of extended object thermal features Co1 vs. So1.	99
4.2	Scatter plot of extended object thermal features Uo1 vs. En1.	100
4.3	Dot plot of extended object thermal feature To1.	100
4.4	Scatter matrix of remaining extended object thermal features after a preliminary feature analysis.	101
4.5	Scatter plot of compact object thermal features Co1 vs. So1.	102
4.6	Scatter plot of compact object thermal features Uo1 vs. En1.	103
4.7	Dot plot of compact object thermal feature To1.	104
4.8	Scatter matrix of remaining compact object thermal features after a preliminary feature analysis.	104
4.9	K-Nearest-Neighbor density estimation.	109
4.10	Principal component analysis used to project patterns onto eigenvector in direction of maximum variance of the patterns.	117
4.11	General trend for extended objects of dotplots with average error rates for each classifier and error estimation method observed in each dimension.	125

xvi   List of Figures

Figure		Page
4.12	Extended object scatter plot of average error rates (%) for KNN classifier (with holdout error estimation method) and KNN classifier (with leave-one-out error estimation method) in three dimensions. Feature vector < 1, 6, 18 > results in the minimum average error rates with the smallest absolute difference in the error rates on the test data set for each error estimation method used by the KNN classifier.	133
4.13	General trend for compact objects of dotplots with average error rates for each classifier and error estimation method observed in each dimension.	140
4.14	Visible images and thermal images for each viewing angle of extended objects used in sensitivity analysis for the variations in the camera's viewing angle. The viewing angles of the thermal images are arranged from left to right as $-60°$ from normal incidence, $-45°$ from normal incidence, $-30°$ from normal incidence, normal incidence, $30°$ from normal incidence, $45°$ from normal incidence, and $60°$ from normal incidence. (a) brick wall, (b) hedges, (c) picket fence, (d) wood wall.	146–147
4.15	Visible images and thermal images for extended objects used in sensitivity analysis for the variations in the window size of the thermal scene. The first (largest) and 100th (smallest) window segments out of the 100 window sizes are enclosed by the solid red borders. (a) brick wall, (b) hedges, (c) picket fence, (d) wood wall.	149
4.16	Brick wall sensitivity analysis for the variations in the window size of the thermal scene. (a) Posterior probabilities for the brick wall feature vectors and (b) macro feature values with variations in window size indexed from 1 (largest window) to 100 (smallest window).	151
4.17	Hedges sensitivity analysis for the variations in the window size of the thermal scene. (a) Posterior probabilities for the hedges feature vectors and (b) macro feature values with variations in window size indexed from 1 (largest window) to 100 (smallest window).	152

Figure		Page
4.18	Picket fence sensitivity analysis for the variations in the window size of the thermal scene. (a) Posterior probabilities for the picket fence feature vectors and (b) macro feature values with variations in window size indexed from 1 (largest window) to 100 (smallest window).	153
4.19	Wood wall sensitivity analysis for the variations in the window size of the thermal scene. (a) Posterior probabilities for the wood wall feature vectors and (b) macro feature values with variations in window size indexed from 1 (largest window) to 100 (smallest window).	154
4.20	Visible image and thermal images for the pine tree log used in the sensitivity analysis for the variations in the rotational orientation. (a) 0°, (b) 45°, (c) 90°, (d) 135°, (e) 180°. The portion of the pine tree log segmented for the analysis is enclosed by the solid rectangular borders in each thermal image.	156
5.1	First principal eigenvectors each projected through the hyperconoidal cluster of their respective object class in a 3-dimensional feature space.	163
5.2	Distance metrics $comp_{\underline{e}_{1j}} \underline{\tilde{f}}_{ij}$ and D_{ij} used to analyze the behavior of each object class's patterns $\underline{\tilde{f}}_{ij}$ about the respective first principal eigenvector $\underline{e}_{1j}$.	165
5.3 (a–j)	Extended object distance metric relations for given most favorable feature vector.	166–175
5.4 (a–r)	Compact object distance metric relations for given most favorable feature vector.	176–192
5.5	Portion of hyperconoidal clusters presented in Fig. 5.1 with an unknown pattern displayed as the black star in the feature space.	195

xviii   List of Figures

Figure   Page

5.6   Visible and thermal images of extended objects from the training
       data set. The thermal images display the thermal radiance and
       contrast that are typically found in the scenes for each object
       class and reference emitters in their respective training data set.
       (a) brick wall (b) hedges, (c) picket fence, and (d) wood wall.   211

5.7   Visible and thermal image of brick wall from the blind data set
       that was misclassified as a hedge by the adaptive Bayesian
       classifier. The thermal image was captured on 24 September
       2007 at 1005 hrs.   212

5.8   Visible and thermal image of hedges from the blind data set that
       was misclassified as a brick wall by the adaptive Bayesian
       Classifier. The thermal image was captured on 15 August 2007
       at 1048 hrs.   214

5.9   Visible and thermal images of a picket fence from the blind data
       set that was misclassified as a wood wall by the adaptive Bayesian
       Classifier. The thermal image was captured on 6 October 2007 at
       1240 hrs.   215

5.10   Visible and thermal images of wood walls from the blind data
       set that were misclassified by the adaptive Bayesian Classifier.
       (a) misclassified as a brick wall (captured on 15 August 2007 at
       1034 hrs), (b) misclassified as a picket fence (captured on
       24 September 2007 at 1029 hrs, same object as in (c) but
       viewed at normal incidence), (c) misclassified as hedges
       (captured on 24 September 2007 at 1030 hrs, same object as in
       (b) but at 45 degrees from normal viewing angle).   216

5.11   Visible and thermal images of compact objects from the training
       data set. The thermal images display the thermal radiance and
       contrast that are typically found in the scenes for each object
       class and reference emitters in their respective training data
       set. Steel poles: (a) brown painted surface, (b) green painted
       surface, (c) octagon shape, w/ aged brown painted surface.
       Tree: (d) basswood tree, (e) birch tree, (f) cedar tree.   218

5.12   Visible and thermal images of a steel pole from the blind data set
       that was misclassified as a tree by the adaptive Bayesian Classifier.
       The thermal image was captured on 5 November 2007 at 1428 hrs.   220

List of Figures xix

Figure		Page
5.13	Visible and thermal images of a tree from the blind data set that was misclassified as a steel pole by the adaptive Bayesian Classifier. The thermal image was captured on 18 September 2007 at 1407 hrs.	221
5.14	Adaptive Bayesian Classification Model Algorithm.	224
5.15	Visible and thermal images of extended blind objects that include classes outside the given training data set. (a) brick wall with moss on the surface, (b) concrete wall, (c) bush, (d) gravel pile, (e) steel picket fence, (f) wood bench, and (g) wood wall of a storage shed.	248
5.16	Visible and thermal images of compact blind objects that include classes outside the given training data set. (a) square steel pole, (b) aluminum pole for dryer vent, (c) concrete pole, (d) knotty tree, (e) telephone pole, (f) 4×4 wood pole, and (g) pumpkin.	249
6.1	(a) visible image, (b) thermal images, (c) frequency spectrum, and (d) polar spectrum of a wood wall.	261
6.2	(a) visible image, (b) thermal images, (c) frequency spectrum, and (d) polar spectrum of a brick wall.	262
6.3	Scaled frequency energy histograms: (a) wood wall and (b) brick wall.	264
6.4	Bayesian multi-sensor data fusion architecture involving thermal infrared and sonar sensors.	265
6.5	Autonomous robot estimates prior probabilities of objects in area of operation using satellite imagery to assist in classifying objects within field-of-view of onboard sensors.	267

List of Tables

Table		Page
2.1	Procedure to normalize the camera and store the reference in the camera's memory to perform non-uniformity correction on subsequent thermal image frames [5].	32
2.2	Procedure to disable AGC by making modifications in the *Raytheon ControlIR 2000B*'s memory using the *Control IR Manager* software [5].	35
2.3	Distribution of training and test data collected from 15 March to 3 July 2007.	42
2.4	Distribution of blind data collected from 6 July to 5 November 2007.	44
3.1	Thermal image capture times and temperatures for objects in Fig. 3.9 captured on 10 February 2007.	67
3.2	Feature values generated from the thermal image of objects in Fig. 3.9 captured on 10 February 2007.	68
3.3	Summary of meteorological, micro, and macro features.	86
3.4	Feature values generated from the thermal image of objects in Fig. 3.17.	87
3.5	Curvature Algorithm used to distinguish compact and extended objects.	89
3.6	Curvature Algorithm demonstration results using objects in Fig. 3.17.	91

Table		Page
4.1	Confusion matrix example that assesses a classification model's performance on test data set consisting of extended objects.	121
4.2	Extended object thermal features and labels used in the exhaustive search feature selection method.	124
4.3	Total number of extended object thermal feature combinations for feature vectors from 1 to 18 dimensions. The first 11 dimensions (highlighted in yellow) satisfy the rule of thumb to ensure peak performance of the classification models.	125
4.4 (a–e)	Extended object comparison of the lowest average error rates (%) of each classifier with the respective error estimation method across each feature vector dimension.	126–128
4.5 (a–b)	Extended object candidates for most favorable feature vectors.	128–129
4.6	Extended object set of most favorable feature vectors for each classifier with the respective error estimation method.	129
4.7 (a–c)	Extended object comparison of the lowest average error rates (%) for combinations of a classifier and error estimation methods across each feature vector dimension.	131–132
4.8	Extended object set of most favorable feature vectors for combinations of a classifier and error estimation methods.	132
4.9	Extended object set of most favorable feature vectors (combined feature vectors from Tables 4.6 and 4.8).	132
4.10	Compact object thermal features and labels used in the exhaustive search feature selection method.	134
4.11	Total number of compact object thermal feature combinations for feature vectors from 1 to 15 dimensions. All 15 dimensions (highlighted in yellow) satisfy the rule of thumb to ensure peak performance of the classification models.	135
4.12 (a–d)	Compact object comparison of the lowest average error rates (%) of each classifier with the respective error estimation method across each feature vector dimension.	136–137

List of Tables xxiii

Table		Page
4.13 (a–c)	Compact object candidates for most favorable feature vectors.	138–139
4.14	Compact object set of most favorable feature vectors for each classifier with the respective error estimation method.	139
4.15 (a–c)	Compact object comparison of the lowest average error rates (%) for combinations of a classifier and error estimation methods across each feature vector dimension.	140–141
4.16	Compact object set of most favorable feature vectors for combinations of a classifier and error estimation methods.	142
4.17 (a–b)	Compact object set of most favorable feature vectors (combined feature vectors from Tables 4.14 and 4.16).	142–143
4.18	Variations in the camera's viewing angle effect on feature values and classification performance of a Bayesian classifier for each extended object in the left column. The object class assigned by the classifier as well as the posterior probabilities for each object class is presented in the columns on the right.	145
4.19	Effect of variations in the rotational orientation on feature values and classification performance of a Bayesian classifier of a pine tree log. The object class assigned by the classifier as well as the posterior probabilities for each rotation angle is presented in the columns on the right.	155
5.1	Comparison of average error rates (%) for adaptive Bayesian classifiers with KNN and Parzen classifiers using most favorable feature vectors and blind data for extended objects. The table cells with the lowest average error rates for each classifier are shaded in gold. The table cell with the overall lowest average error rate is shaded in green.	198
5.2 (a–b)	Comparison of average error rates (%) for adaptive Bayesian classifiers with KNN and Parzen classifiers using most favorable feature vectors and blind data for compact objects. The table cells with the lowest average error rates for each classifier are shaded in gold. The table cells with the overall lowest average error rate are shaded in green.	199

Table		Page
5.3a	Brick wall lowest error rates with respective feature vector and distance function combination displayed in the upper left corner of each confusion matrix.	200–201
5.3b	Hedges lowest error rates with respective feature vector and distance function combination displayed in the upper left corner of each confusion matrix.	202–203
5.3c	Picket fence lowest error rates with respective feature vector and distance function combination displayed in the upper left corner of each confusion matrix.	204–205
5.3d	Wood wall lowest error rates with respective feature vector and distance function combination displayed in the upper left corner of each confusion matrix.	206–207
5.4 (a–b)	Steel Pole and Tree lowest error rates with respective feature vector and distance function combination displayed in the upper left corner of each confusion matrix.	208–209
5.5 (a–d)	Confusion matrices of the Adaptive Bayesian Classification Model with various threshold values for the extended objects. Fixed threshold values are noted in the upper left corner. Threshold with a varied value is noted at the upper left corner of each matrix. Thresholds highlighted in green colored text are selected as most favorable for the Adaptive Bayesian Classification Model applied to the extended objects.	228–235
5.6 (a–d)	Confusion matrices of the Adaptive Bayesian Classification Model with various threshold values for the compact objects. Fixed threshold values are noted in the upper left corner. Threshold with a varied value is noted at the upper left corner of each matrix. Thresholds highlighted in green colored text are selected as most favorable for the Adaptive Bayesian Classification Model applied to the compact objects.	236–243

Table		Page
5.7 (a–d)	Comparison of confusion matrices of the best performing classification models applied to the extended objects from the Adaptive Bayesian Classification Model (via Committees of Experts), Adaptive Bayesian Classifier with single distance function, KNN Classifier, and Parzen Classifier.	244–245
5.8 (a–d)	Comparison of confusion matrices of the best performing classification models applied to the compact objects from the Adaptive Bayesian Classification Model (via Committees of Experts), Adaptive Bayesian Classifier with single distance function, KNN Classifier, and Parzen Classifier.	246–247
5.9	(a) Adaptive Bayesian Classification Model class assignments and posterior probabilities on extended blind objects displayed in Fig. 5.15. (b) Threshold values for the Adaptive Bayesian Classification Model.	250
5.10	(a) Adaptive Bayesian Classification Model class assignments and posterior probabilities on compact blind objects displayed in Fig. 5.16. (b) Threshold values for the Adaptive Bayesian Classification Model.	250

Contents

1 Introduction and Overview 1
   1.1 Purpose of Book 1
   1.2 Non-Heat Generating Objects 6
   1.3 Autonomous Robotic Systems 8
       1.3.1 Detect the Object 11
       1.3.2 Segment the Object 13
       1.3.3 Classify the Object 13
   1.4 Infrared Thermography 15
       1.4.1 Active vs. Passive Thermography 16
       1.4.2 Advantages & Disadvantages
              of Thermal Infrared Imaging 17
       1.4.3 Multi-Mode Heat Transfer Model 18
   1.5 Overview of the Book 20
   References 22

2 Data Acquisition 25
   2.1 Introduction 25
   2.2 Robotic Thermal Imaging System 25
       2.2.1 Hardware 25
       2.2.2 Signal Preprocessing 28
   2.3 Data Collection 41
   2.4 Summary 45
   References 45

3 Thermal Feature Generation 47
   3.1 Introduction 47
   3.2 "Ugly Duckling" Features 48
   3.3 Thermal Image Representation 51

	3.4	Meteorological Features	54
		3.4.1 Ambient Temperature	54
		3.4.2 Ambient Temperature Rate of Change	54
	3.5	Micro Features	55
		3.5.1 Emissivity Variation by Material Type	55
		3.5.2 Emissivity Variation by Viewing Angle	58
		3.5.3 Emissivity Variation by Surface Quality	59
		3.5.4 Emissivity Variation by Shape and Surface Temperature	59
		3.5.5 Other Directional Variation Enhancers	60
		3.5.6 Emissivity-based Features	63
	3.6	Macro Features	69
		3.6.1 First-order Statistical Features	70
		3.6.2 Second-order Statistical Features	74
	3.7	Thermal Feature Application	84
	3.8	Curvature Algorithm	88
	3.9	Summary	91
	References		91
4	**Thermal Feature Selection**		**95**
	4.1	Introduction	95
	4.2	"No Free Lunch" Classifiers	96
	4.3	Preliminary Feature Analysis	98
	4.4	Classifiers	105
		4.4.1 Bayesian Classifier	105
		4.4.2 K-Nearest-Neighbor (KNN) Classifier	110
		4.4.3 Parzen Classifier	111
		4.4.4 General Remarks	113
	4.5	Model Performance and Feature Selection	115
		4.5.1 Feature Selection Method	116
		4.5.2 Performance Criterion	120
		4.5.3 Error Estimation Method	122
		4.5.4 Checkpoint Summary	123
		4.5.5 Extended Object Performance and Feature Selection	123
		4.5.6 Compact Object Performance and Feature Selection	133
	4.6	Sensitivity Analysis	143
		4.6.1 Viewing Angle Variations	144
		4.6.2 Window Size Variations	148
		4.6.3 Rotational Variations	155
	4.7	Summary	156
	References		158

5	**Adaptive Bayesian Classification Model**		161
	5.1	Introduction	161
	5.2	Distance Metrics for Hyperconoidal Clusters	162
	5.3	Adaptive Bayesian Classifier Design	193
	5.4	Adaptive Bayesian Classifier Appraisal	197
		5.4.1 Blind Data Performance	197
		5.4.2 Analysis of Misclassifications	210
	5.5	Adaptive Bayesian Classification Model Design	223
	5.6	Adaptive Bayesian Classification Model Application	227
		5.6.1 Performance on Blind Data (with Classes = Training Set)	228
		5.6.2 Performance on Blind Data (with Classes ≠ Training Set)	247
	5.7	Summary	251
	References		254
6	**Conclusions and Future Research Directions**		255
	6.1	Introduction	255
	6.2	Contributions	256
	6.3	Limitation of a Thermal Infrared Imaging System	257
	6.4	Future Research	259
		6.4.1 Augmentation of Robotic Thermal Imaging System	259
		6.4.2 Fuzzy Logic Classifier	260
		6.4.3 Bayesian Multi-Sensor Data Fusion	265
		6.4.4 Prior Knowledge Based on Satellite Imagery	267
	6.5	Concluding Remarks	268
	References		269
Index			**271**

1 Introduction and Overview

Abstract This chapter introduces the objective of this book, to present the design and implementation of a physics-based pattern classification model to characterize non-heat generating outdoor objects in thermal scenes for autonomous robots. The classification of indoor objects and heat generating objects is a solved problem. However, a missing and essential piece in the literature is research involving the automatic characterization of non-heat generating objects in outdoor environments using a thermal infrared imaging modality for mobile robotic systems. Seeking to classify non-heat generating objects in outdoor environments using a thermal infrared imaging system is a complex problem due to the variation of radiance emitted from the objects as a result of the diurnal cycle of solar energy. The model design cycle outlined for presentation in subsequent chapters will allow bots to "see beyond vision" to autonomously assess the physical nature of the surrounding structures for making decisions without the need for an interpretation by humans.

1.1 Purpose of Book

The goal of the work presented in this book is to complement an autonomous robot's situational awareness by providing the ability to classify smaller structures commonly found in the immediate operational environment. These are structures that cannot be assessed in enough detail by GPS and onboard terrain mapping systems currently configured on bots. Situational awareness is the bot's interpretation of objects and physical processes in its internal representation of the environment. Mobile bots operating independently in unstructured outdoor environments must maintain situational awareness to permit sound decisions. The bot's internal representation of the environment is formed by the synthesis of prior knowledge and information obtained from sensors. The bot develops an interpretation by detecting, segmenting (or distinguishing), and classifying objects and physical processes within its internal representation. Based on this interpretation, the bot can

2   1 Introduction and Overview

decide on how to respond to situations and what actions are necessary to accomplish a given task. Autonomous bots will require the ability to make decisions such as "go through the hedges" or "go around the brick wall." To carry out these types of actions, the bot must have the ability to classify the unknown object as being either hedges or a brick wall. Therefore, our interest is in the situation where the bot has already detected and segmented a non-heat generating object but now needs to classify the object in a highly unstructured outdoor environment like those presented in Fig. 1.1, to include conditions of limited visibility.

We envision mobile bots that unobtrusively travel alongside pedestrians at a walking pace in an unstructured environment. It is important that small mobile

Fig. 1.1 Unstructured environments as potential areas of operation for autonomous robots.

robots, with wheels, legs, and/or tracks, normally travel at the same speed as the pedestrian traffic, even if they traverse to quickly move down a vacant alley to conduct a reconnaissance or slow down to characterize an obstacle, because people resent having to go around a slow bot while they are also startled by machines such as Segways and golf carts that overtake them without warning. Furthermore, the type of sensors used to afford the bot with situational awareness is tied to the speed of the bot. A thermal infrared imaging modality mounted on a mobile robot is a favorable choice for receiving enough detailed information to automatically interpret objects at close ranges relevant to walking speeds. The technology necessary for thermal imaging has just recently become sufficiently portable and inexpensive enough to mount on small robotic platforms. Furthermore, passive thermal infrared imaging modalities do not pose a risk to humans like one might have with laser-based sensors, such as LADAR. Our use of a thermal infrared imaging modality will not only afford the ability to identify targets during conditions of limited visibility but it will also eliminate the need for a light source mounted on a bot to illuminate targets for classification that could disclose the bot's location. For example, illuminating the fence in Fig. 1.2 (a) with a visible light source as in Fig. 1.2 (b) would reveal the tactical position of the bot and perhaps compromise any reconnaissance missions. On the other hand, the thermal infrared imaging system that simultaneously captured the image of the fence in Fig. 1.2 (c) acts as a passive system that does not emit any visible signatures for enemy detection. The thermal infrared imaging sensor is a passive system since there is no need for an onboard artificial illumination source to operate. The only source required for the fence to emit thermal energy is the sun that provides solar energy during the daylight hours.

The objective of this book is to design and implement a physics-based pattern classification model to characterize non-heat generating outdoor objects in thermal scenes for autonomous robots. The classification of indoor objects and heat generating objects is a solved problem. However, a missing and essential piece in the literature is research involving the automatic characterization of non-heat generating objects in outdoor environments using a thermal infrared imaging modality for mobile robotic systems. Seeking to classify non-heat generating objects in outdoor environments using a thermal infrared imaging system is a complex problem due to the variation of radiance emitted from the objects as a result of the diurnal cycle of solar energy. Our desired model will allow bots to "see beyond vision" to autonomously assess the physical nature of the surrounding structures as well as report classes of objects while performing security or reconnaissance missions. We will design a classification model that retains the original physical interpretation of the information in the signal data throughout the classification process. This emphasis will result in a framework that allows the human analyst to understand the reason for a bot's classification of an unknown object by associating the final classification decision with the thermal-physical properties found in the original signal data. Additionally, our approach will afford bots with the intelligence to automatically interpret the information in signal data to make decisions without the need for an interpretation by humans.

4   1 Introduction and Overview

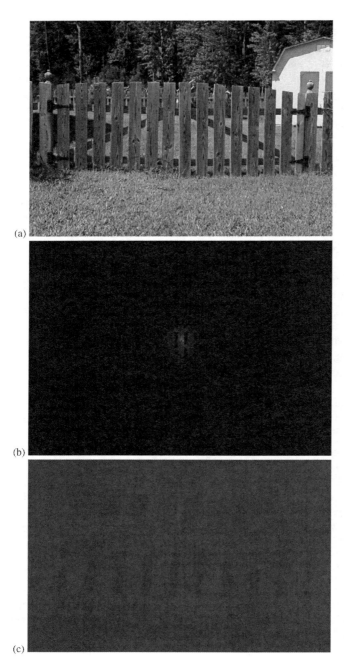

Fig. 1.2 Visible and thermal images of a wooden fence. (a) visible image of the fence during the day, (b) visible image captured at 2030 hrs on 7 September 2007 with light source illuminating on the fence, (c) thermal image of the fence captured at the same time as the visible image in (b) and at an ambient temperature of $71.9\,°F$.

The research presented in this book evolved from a broader work, by the Nondestructive Evaluation Laboratory at The College of William & Mary, to automate the fusion and interpretation of data streams from various active and passive sensor systems to enable autonomous mobile robot operations in a wide variety of unstructured outdoor environments. We feel that it is the fusion of an active sensor, such as sonar (air-coupled ultrasound), and a passive sensor, such as thermal infrared and RGB video, systems that has the potential for the greatest advancements because of the complementary nature of the modalities. Two mobile robots, displayed in Fig. 1.3, are currently being used to collect systematic ultrasonic and infrared imagery data streams about The College of William & Mary campus, the adjacent colonial area, York County, Virginia, in a village and on a farm outside of Buffalo, New York, and on mountainous terrain in Eleanor, West Virginia. We have used these large data sets to explore the behavior of features generated from the signal data of classes of outdoor objects and design single-sensor classification algorithms that afford mobile robots the ability characterize outdoor objects. The research presented in this book is an extension to our previous work involving sonar sensor interpretation by mobile robots [1]. This research involves the design of algorithms to distinguish outdoor objects such as trees, poles, fences, walls, and hedges based on features generated from backscattered sonar echoes. Our novel model involving thermal infrared imagery presented in Chap. 5 of this book affords a complementary technique to classify the same types of objects. Since both ultrasound and infrared are independent of lighting conditions, they are appropri-

Fig. 1.3 Mobile robotic 3D sonar scanning system, *rWilliam* (on right) and thermal imaging system, *rMary* (on left).

ate for use both day and night. In Chap. 6, we will discuss our future research that is aimed towards designing a framework that fuses information from the bot's thermal infrared imaging and ultrasonic sensors to perform intelligent actions, such as decision-making and learning.

1.2 Non-Heat Generating Objects

Non-heat generating objects are defined as objects that are not a source for their own emission of thermal energy, and so exclude people, animals, vehicles, etc. Non-heat generating objects can be natural or human-made. Our choices of natural objects that do not generate their own thermal energy include trees and bushes. Human-made objects include brick walls, wood walls, fences, and steel poles. Consequently, the ability of non-heat generating objects to display a thermal signature depends partly on the thermal energy received from heat generating sources in the environment. The primary heat generating source is the sun. However, there may also exist other objects in the local environment that generate and emit their own thermal energy and/or reflect thermal energy emitted from other sources. The ability for a non-heat generating object to display a thermal signa-

Fig. 1.4 Thermal scene consisting of heat and non-heat generating objects. Heat generating objects include the human walking on the sidewalk and squirrel running from behind the tree. Non-heat generating objects include the trees and steel pole used by the street light.

1.2 Non-Heat Generating Objects   7

ture also depends on its physical composition. We will discuss the thermal emission characteristics of non-heat generating objects in Chap. 3.

Identifying heat generating objects in thermal scenes, using pattern classification techniques, has become relatively trivial because infrared imaging cameras are very sensitive to detecting the thermal contrast between the object and surrounding surfaces. For instance, the human walking on the sidewalk and squirrel running from behind the tree in Fig. 1.4 can be identified by generating geometric features from various points on the body such as those presented in Fig. 1.5 (a). Features are unique representations of an object class that are generated from an object's signal received by a sensor. These features are used by a pattern classification model to distinguish one object class from another and provide class assignments to unknown objects. Geometric features can also be generated from tires and different segments of vehicle surfaces for class assignments as displayed in Fig. 1.5 (b). However, generating features from the thermal image of a non-heat generating object like the trees and steel poles in Fig. 1.4 for classification is

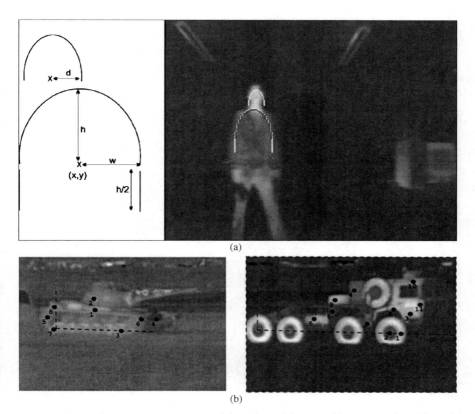

Fig. 1.5 Geometric measurements generated from thermal images of heat generating objects for classification. (a) measurements generated to classify people [2]. (b) measurements generated to classify vehicles [3].

8   1 Introduction and Overview

a more subtle process due to the variation in thermal radiance of objects in the scene primarily caused by the diurnal cycle of solar energy. We will provide a detailed discussion on techniques used to generate features for heat and non-heat generating objects in Chap. 3 and present various classifiers used in classification models in Chap. 4. In Chap. 5, we will present our novel classification model that outperforms the traditional classifiers when characterizing non-heat generating objects in outdoor environments.

1.3 Autonomous Robotic Systems

Robots have many uses in the military, industry, health care services, and neighborhood homes. A general summary of the current uses of robots is provided in [4]. Robots categorized as unmanned ground, marine, and aerial vehicles are normally found in the military. In industry, robots are commonly used on assembly lines in automotive and food processing plants. These robots are usually in the category of machine vision and used to assemble products and/or detect defects in the products. In health care, robots are now used to assist during surgical procedures. Robotic devices are also starting to be used to assist elderly people, particularly in Japan. We can also find robots in homes in the form of vacuum clean-

Fig. 1.6 Roomba vacuum cleaning robot.

ers and even lawn mowers. Each type of robot operates at specific level of autonomy. The level of autonomy afforded to robots usually depends on the size and mobility capabilities of the bot and level of risk in harming humans and pets. Though the Roomba vacuum cleaners in Fig. 1.6 are semi-autonomous, we would have no problem with letting them roam anywhere around the house since the bots are ankle high. On the other hand, we would expect the robotic lawn mower in Fig. 1.7 to have a higher level of intelligence so the neighbor's favorite tulips are not misclassified as a blade of grass. Our objective is to design the intelligence algorithms required by mobile autonomous bots to correctly make decisions regarding non-heat generating objects that exist in their path.

A mobile autonomous robotic system is a ground, marine, or aerial vehicle consisting of all the integrated components (mobility platform, sensors, computers, and algorithms) required to perceive, learn, and adapt in the environment to make intelligent decisions for navigating, communicating, and accomplishing required tasks. A historical background on advances in the state of the art for unmanned ground vehicles from 1959 to 2002 is presented in [5]. The focus of our research is to support autonomous unmanned ground vehicles; however, the framework of our classification model presented in Chap. 5 could be applied to marine and aerial vehicle applications as well.

The robotic platform design is not an issue anymore. Whether the robot will serve the military or be a part of the civilian workforce, the platform will be de-

Fig. 1.7 Robotic lawn mower.

signed to support the required application. For instance, Fig. 1.8 (a) presents a robotic platform that could be used for military reconnaissance missions, Fig. 1.8 (b) shows a robotic platform designed for ambulatory applications, and Fig. 1.8 (c) shows a robotic platform designed for monitoring and surveillance applications. However, the greatest challenge is how to design the intelligence software that will allow the bot to use relevant sensors to learn and make decisions. We obviously hope that the autonomous military reconnaissance vehicle would make the correct classification and decision to go through hedges and not a misclassification that results in the bot attempting to go through a six meter high brick wall. Furthermore, we would expect that an unmanned ambulatory vehicle will extract injured personnel from a burning building and not garbage cans due to misclassifications.

Analogous to living organisms using their senses to understand the environment, autonomous bots will have to interpret information received by their sensors

Fig. 1.8 Autonomous unmanned ground vehicle platforms designed to support various military and commercial applications. (a) military reconnaissance application (Courtesy of globalsecurity.org), (b) Battlefield Extraction Assist Robot for ambulatory applications (Courtesy of Vecna Technologies, Inc.), (c) remote monitoring and surveillance applications (Courtesy of MobileRobots, Inc.).

to detect, segment, and classify natural and human-made objects. Sensors used to detect, segment, and classify objects are either active or passive sensors. Active sensors require an external or onboard source to transmit a signal that is reflected by the target and then received by the bot's sensor. Passive sensors do not require an active onboard source to transmit energy at a target. Thus, passive sensors receive signal information that is naturally emitted from an object's surface. Detection involves comparing signals received within a sensor's field of view to determine whether an object is present. Once detected the object is segmented to distinguish it from the surrounding environment. The segmented object is then assigned to a specific object class based on the bot's assessment of the object and previous knowledge about the local area of operation. The autonomous bot can then make a decision pertaining to the classified object depending on the required task or mission. For instance, if the object is a trash can, the bot may be required to report the trash can and quietly go around it when on a reconnaissance mission or pick it up and empty the can in the dumpster when performing janitorial duties. In any case, the autonomous bot must have the intelligence to classify non-heat generating objects.

1.3.1 Detect the Object

Detection of obstacles by bots is quite trivial nowadays. For instance, with an active sensor system, a source simply transmits some pulse of energy from the robot's platform and onboard sensors receive the energy after being reflected from an object in the path. The bot's intelligence software analyzes contrasting information in the reflected signals received within the field of view of the sensor to determine the ranges, sizes, and locations of objects. Consequently, detection usually coincides with obstacle avoidance. Thus, the bot simply knows the location and size of an unknown object in its path and travels around the object to avoid a collision. The Defense Advanced Research Projects Agency (DARPA) Grand Challenge, that took place in the Mojave Desert of southwestern United States on 8 October 2005, proved that sophisticated semi-autonomous robots are able navigate along a grueling route by using multiple sensors to detect obstacles and map the terrain [www.darpa.mil]. Active sensors normally used by bots to detect objects include laser detection and ranging (LADAR), synthetic aperture radar (SAR), ultrasound, and infrared sensors. An advantage of LADAR is that it has exceptional resolution; however, a disadvantage is that it is affected by dust and smoke that may be interpreted as an object in the bot's path [5]. Additionally, certain tactical situations may limit the use of LADAR due to its potential risks to humans. Although SAR performs well in the presence of obscurants, it lacks spatial resolution and may not detect non-metallic objects depending on their moisture content [5]. Ultrasound transducers display exceptional performance in detecting objects during conditions of limited visibility and in the presence of obscurants such as dust, smoke, and fog at short ranges. Furthermore, ultrasound does not have any safety concerns like those associated with LADAR. An example of how ultrasound sensors can be used to detect and avoid obstacles is given in [6]. An infrared sensor performing in an active role requires a transmitter to emit energy at an object and the sensor to receive the energy reflected from the object's surface. For instance, the infrared detection and range sensor system in Fig. 1.9 transmits a pulse of infrared energy from an emitter that is a fixed distance from the detector. If the energy hits an object, reflected waves are received by a specific portion of a linear charge-coupled device (CCD) array in the detector based on the angle of the wave. The angles in the triangle formed by the emitter, point of reflection, and detector vary based on the distance to the object. Thus, the sensor uses the reflected wave's point of impact on the CCD array to complete the triangle and estimate the distance to the object. A method for detecting and estimating distances to objects using ultrasound and active infrared sensors is discussed in [7]. An emerging active sensor that operates at 110 GHz to 10 THz, between microwaves and the infrared bands, in the electromagnetic spectrum involves terahertz-pulsed imaging. Research interests using terahertz-pulsed imaging involve applications such as detection of concealed weapons and explosives [8]. An advantage of using terahertz radiation for these applications is that metals are opaque to the radiation. Additionally, terahertz radiation poses no health risk to humans. A limiting factor is that most non-metals, such as non-heat generating wooden fences, are transparent to terahertz and propagation distance is limited at the higher frequencies. However, this limitation could be abated by the terahertz band's sensitivity to the presence of water, which may be of use for not only detecting (and characterizing) the disease states of human tissue [9] but also other living objects such as trees and bushes.

Passive sensors include red, green, blue (RGB) vision cameras and thermal infrared detectors. RGB cameras provide excellent resolution but are limited to operation during times when no obscurants are present and the target is illuminated

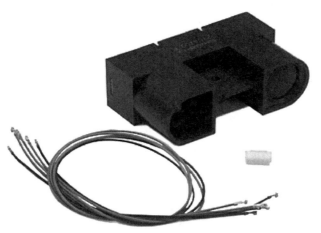

Fig. 1.9 Infrared range sensor with detection range from 1 to 5.5 m (Courtesy of Acroname, Inc.)

with light. In a passive role, the infrared sensor is usually a focal plane array (FPA) of thermal (or long-wave) infrared detectors that operate at 7 to 14 μm in the electromagnetic spectrum. Unlike the 1-dimensional array used by the active infrared range sensor, the passive thermal infrared sensor consists of a 2-dimensional FPA of detectors. Thermal radiance emitted by an object and received on the FPA is converted to an analog signal. This analog signal is then converted to a digital signal for display as a thermal image. Objects are detected using the thermal infrared imaging sensor by analyzing thermal contrasts in the signal information received passively from the surrounding environment within the field of view of the sensor. A comparison of thermal infrared detection algorithms is found in [10]. Since the thermal infrared imaging sensor is used in our current research, we will delay our discussions of the characteristics of this sensor until Sect. 1.4 and subsequent chapters in this book. To complement a bot's ability to detect objects, the intelligence algorithm normally uses more than one type of sensor. Object detection and avoidance methods using vision and ultrasonic sensors for mobile bots are discussed in [11, 12, 13]. A technique for detecting objects using ultrasound and passive infrared sensors is discussed in [14].

1.3.2 Segment the Object

Once a target is detected by displaying a signal difference from other objects in the sensor's field of view, it is segmented from its surroundings and prepared for classification by the bot's intelligence algorithm. Discussions on the detection and segmentation of objects in infrared images are found in [15, 16]. Techniques for segmentation of objects in general images are discussed in [17]. Preparing the segmented signal information for the classification phase involves preprocessing to minimize the effects of temporal and spatial signal degradations. The preprocessing must avoid the use of filters that would lead to loss of relevant signal information used in the classification phase. We provide a detailed discussion on acquisition and preprocessing of thermal infrared images in Chap. 2.

1.3.3 Classify the Object

After segmenting and preprocessing the unknown object, the bot uses its intelligence algorithms to classify the object. The autonomous bot can then make a decision pertaining to the classified object depending on the required task or mission. The design of the classification model continues to be the most challenging phase for any intelligence system. In this research we will assume that the bot has already detected and segmented an unknown object. Therefore, our objective is to design and implement a model that will allow the bot to classify the unknown object. Two approaches can be used to design a model that will assign a class to an unknown object – theoretical models (analytical or numerical) and observational models. Theoretical models normally involve the use of differential equations to compute the estimated value of physical variables associated with unknown objects for comparison with measured values from known objects. Class assignment is determined by the computed values' closeness to the measured values. Theoretical models usually require at least one measured value for the parameters in the model. These measured values are obtained by using an instrument that makes contact with the object. One possibility for obtaining physical measurements from an unknown object is by equipping a bot with touch sensors [18]. However, a bot that can classify objects using non-contact sensors is more practicable. We will continue our discussion on a specific theoretical model known as the multi-mode heat transfer model in Sect. 1.4.

Our method of choice for designing a classification algorithm is the observational model approach. An observational model estimates class assignments of unknown objects based on inferences made from empirical knowledge and prior knowledge. The empirical knowledge is obtained by observing information received by the sensors. The prior knowledge is based on observations regarding the presence of objects existing in the bot's area of operation before entering the area. The empirical knowledge and prior knowledge are combined to produce posterior knowledge that yields a class assignment for the unknown object. Observational models are used in the field of pattern classification (or recognition). Pattern classification is the process of characterizing an unknown object based on an assessment of attributes (also called features or patterns) that are generated from the object's signal received by a sensor. The class assignment of the unknown pattern is made by a classification model consisting of a classifier and features that uniquely represent each object class requiring classification. The success of a classification model relies primarily on the selection of features that provide the most favorable distinction between each object class. However, a poor choice of feature types and/or generating features that are not representative of objects in the bot's area of operations will result in ambiguity with separation of object classes and ultimately an increase in the misclassification rate. We will provide a detailed discussion on choices for features and approaches for pattern recognition in Chaps. 3 and 4, respectively. While designing our classification model, presented in Chap. 5, we will make considerable effort to provide guidance on how to analyze features to understand their underlying physics and select most favorable sets of features that minimize the misclassification of unknown objects. Additionally, our approaches to feature selection and classification will retain the original physical interpretation of the information in the signal data throughout the classification process.

Our classification of non-heat generating objects (brick walls, hedges, picket fences, wood walls, steel poles, and trees) in outdoor environments could be placed in the category of terrain classification. There are many approaches found in the literature that effectively use various sensors to classify objects in outdoor environments. The design of algorithms to distinguish outdoor objects such as trees, poles, fences, walls, and hedges based on features generated from backscattered sonar echoes for interpretation by mobile robots is discussed in [1]. Discussions on LADAR sensors and object recognition approaches using 3-dimensional LADAR and SAR imagery are presented in [9]. Terrain classification using LADAR to distinguish surfaces (ground surface, rocks, large tree trunk), linear structures (wires, thin branches, small tree trunks), and porous volumes (foliage, grass) for autonomous robot navigation is discussed in [19]. Terrain classification methods using a color vision camera and LADAR to discriminate between soil, vegetation, tree trunks, and rocks for autonomous off-road navigation is presented in [20]. A method for terrain classification involving inertial, motor, ultrasonic, active infrared, microphone, and wheel encoder sensors to classify gravel, sand, asphalt, grass, and dirt is discussed in [21]. The ultrasonic and infrared range sensors were mounted on the robotic platform and aimed downward to the ground to classify the terrain based on the periodogram of the reflected signal (in the frequency domain) and range signal (in the time domain).

The LADAR, SAR, sonar, terahertz-pulsed imaging, and RGB vision modalities presented above all have the capability to complement a bot's intelligence algorithm that is designed to classify objects at close ranges (~2–3 meters) relevant to walking speeds. A thermal infrared imaging modality mounted on a mobile robot is also a favorable choice for receiving enough detailed information to automatically interpret objects at close ranges relevant to walking speeds. However, as we will further discuss in Chap. 3, a missing and essential piece in the literature is research involving the automatic characterization of non-heat generating objects in outdoor environments using a thermal infrared imaging modality for mobile robotic systems. Seeking to classify non-heat generating objects in outdoor environments using a thermal infrared imaging system is a complex problem due to the variation of radiance emitted from the objects as a result of the diurnal cycle of solar energy. Our approach of using a thermal infrared imaging camera for pattern classification makes use of concepts found in the fields of nondestructive evaluation, remote sensing, and digital image processing. Our novel classification model will provide an approach that can make use of thermal infrared imagery as a stand-alone sensor or in combination with other existing sensors to complement the intelligence of a bot. Additionally, the framework of our classification model could also be used in other applications requiring the characterization of unknown objects based on features that witness variations due to natural cyclic events. A somewhat more speculative extension would be an application to autonomous Lunar or Martian rovers, since the diurnal heating effects that we are exploring do not require an atmosphere. On the other hand, ultrasound sensors would not support applications in this environment since nobody can hear you "scream" on the moon or Mars.

1.4 Infrared Thermography

Thermography is the study of internal and/or surface heat distributions of a structure using various instruments that measure thermal energy. Such instruments could require contact techniques such as a probe to measure surface temperatures on the structure. On the other hand, non-contact techniques afford the ability to study heat distributions by measuring the thermal radiation emitted from the surface of the structure using an infrared detector. These noninvasive techniques are used in infrared thermography, which is the foundation for our research presented in this book.

The techniques of infrared thermography are used in the field of nondestructive evaluation (NDE) or thermographic nondestructive testing (TNDT or NDT) to noninvasively assess the behavior of what is at the subsurface of an object. Infrared thermography is widely used in NDE to examine the nature of objects for suitability and quality. Applications are found in areas of preventive maintenance for aircraft (to include space launch vehicles), electrical utilities, and building construction [22, 23]. Applications involving infrared thermography in NDE are also being researched in the field of medicine [24]. Infrared thermography is also used in surveillance operations involving the military, law enforcement, and search and rescue [23].

The applications mentioned above normally require a human operator to assess the thermal image of an object. As we will discuss in Chap. 3, many techniques exist that use pattern recognition methods to automatically classify a target without the need for a human operator. In the military, these approaches are normally referred to as automatic target recognition (ATR) algorithms. However, the majority of the methods available in the literature, using thermal infrared imaging to classify objects, involve heat generating targets. The only use of thermal infrared imaging to classify non-heat generating objects in outdoor environments was found in the area of remote sensing to discriminate between vegetation and soil. We have not identified any previous research in the literature involving the assignment of classes to non-heat generating objects in outdoor environments using a thermal infrared imaging sensor for autonomous robotic systems.

1.4.1 Active vs. Passive Thermography

Analogous to the active and passive functions that the sensors described in Sect. 1.3 have, a thermal infrared imaging system can have either an active or passive role. As mentioned previously, active systems have an external or onboard source to transmit signal energy that is reflected by the target and then received by the sensor. In active thermal infrared imaging, thermal energy from a source is directed towards the specimen being inspected to create differences in the thermal image that identify anomalies in the structure and/or analyze the diffusion of thermal waves to estimate the physical properties of the material. The active heat source used to estimate thermal properties of a given material are formally the boundary conditions that we will present in Sect. 1.4.3 involving the heat transfer model. Methods used to stimulate a specimen with an external source include pulsed thermography, step heating, lock-in thermography, and vibrothermography [22].

Passive thermography does not require an active source to transmit thermal energy at a target. Thus, passive thermal infrared imaging sensors receive thermal radiance that is naturally emitted from an object's surface. The research presented in this book uses passive thermal infrared thermography where the only mandatory source of thermal energy is the sun that provides solar energy during the daylight hours.

1.4.2 Advantages & Disadvantages of Thermal Infrared Imaging

Every sensor has its own advantages and disadvantages. A major advantage of using a thermal infrared imaging sensor is that it provides the ability to identify objects during conditions of limited visibility. Conditions of limited visibility such as night and the presence of obscurants (smoke, light dust, and light haze) have a minimal attenuating effect on long-wave infrared waves. Our choice of a thermal (long-wave) infrared detector yields an operating band of 7 to 14 μm in the electromagnetic spectrum. Long-wave infrared has an advantage over the other bands in the infrared region: near infrared (0.7–1.1 μm), short-wave infrared (1.1–2.5 μm), and mid-wave infrared (2.5–7.0 μm). Figure 1.10 displays the spectral radiance of a perfect emitter of thermal radiation (blackbody) across a band of wavelengths in the electromagnetic spectrum and at various surface temperatures of the blackbody as described by Planck's law. As we see, the long-wave infrared band (denoted by the blue shaded region)

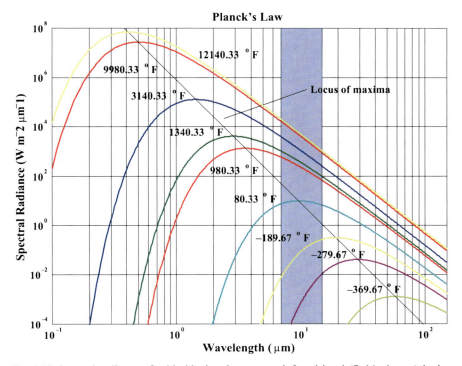

Fig. 1.10 Spectral radiance of a blackbody. Long-wave infrared band (7–14 microns) is denoted by the blue shaded region.

yields the highest thermal radiance for the range of ambient and non-heat generating object surface temperatures encountered by an autonomous mobile robotic system on Earth. Consequently, a thermal infrared imaging camera will maximize the detection of thermal radiance emitted by an object compared to detectors that operate in the near, short-wave, and mid-wave infrared spectral bands.

We will discuss more details of the limitation of using a thermal infrared imaging sensor in Chap. 6. However, we will note a few disadvantages of using this sensor right now. A minor disadvantage is that the thermal infrared imaging camera cannot discriminate between the radiance detected at each wavelength. Thus, in contrast to how the human eye can distinguish the colors red and blue, the thermal infrared imaging camera only "sees" a total radiance from the entire long-wave band of wavelengths. However, this deficiency is tolerated for our application since the FPA of detectors in the thermal infrared imaging camera receives different levels of radiance across the 2-dimensional array to yield a thermal image with related gray-level values. We will discuss the characteristics of the thermal infrared imaging camera in Chaps. 2 and 3.

Since our application takes place outdoors, environmental conditions will exist where the surfaces of a target and surrounding objects will emit approximately the same level of thermal radiance. This phenomenon, known as thermal crossover [23], results in minimal thermal contrast between the surfaces of objects and the surrounding environment within the thermal infrared camera's field of view. Thus, these periods of thermal crossover could result in a limitation in our ability to classify non-heat generating objects in an outdoor environment using a thermal imaging sensor. We will revisit the phenomenon of thermal crossover again in Chaps. 4, 5, and 6.

One possible critical disadvantage of using a thermal infrared imaging camera for autonomous mobile robotic applications is that glass is opaque to infrared radiation. Consequently, a bot will not be able to detect objects that are behind glass. We will revisit this ability of objects to emit thermal energy when we discuss the thermal property known as *emissivity* in Chap. 3.

The disadvantages found with any sensor obviously provide the reason why multi-sensor data fusion systems are normally more successful in classification applications than systems with a single sensor. Thus, the interpretations of relevant information received by different types of sensors used in a multi-sensor framework are fused to complement the overall performance of the classification process. We will discuss our plans for integrating our current pattern classification model using thermal infrared imagery into a multi-senor data fusion framework in Chap. 6.

1.4.3 Multi-Mode Heat Transfer Model

A multi-mode heat transfer equation is a differential equation, along with the corresponding initial and boundary conditions, that models the flow of heat energy by

1.4 Infrared Thermography

conduction, convection, and radiation. Thus, the multi-mode heat transfer equation is a theoretical model. The governing multi-mode heat transfer model for an anisotropic object with no internal heat source is given as [25]:

$$\frac{\partial}{\partial x}\left(K_x \frac{\partial T}{\partial x}\right) + \frac{\partial}{\partial y}\left(K_y \frac{\partial T}{\partial y}\right) + \frac{\partial}{\partial z}\left(K_z \frac{\partial T}{\partial z}\right) = \rho C \frac{\partial T}{\partial t} + \rho C t_r \frac{\partial^2 T}{\partial t^2} \quad (1.1)$$

$$T(t=0) = T_0 \quad (1.2)$$

$$-K_n \frac{\partial T}{\partial \underline{n}} = Q - h_c(T - T_a) - \varepsilon \sigma \left(T^4 - T_a^4\right) \quad (1.3)$$

$$T_s = T_d \quad (1.4)$$

$$Q_s^{cd} = Q_d^{cd} + Q_d^{cn} + Q_d^{r} \quad (1.5)$$

where T is the temperature of the object and T_a is the ambient temperature; ρ and C are the density and specific heat of the object, respectively; K_x, K_y, and K_z are the in-plane and transverse thermal conductivity of the object; t_r is the relaxation time; $\underline{n}$ is the vector normal to the object's surface; Q is the heating flux; h_c is the free convection coefficient; ε is the object's emissivity; σ is the Stephan-Boltzmann coefficient; indices s and d specify the object specimen and defect, respectively; and indices cd, cn, and r specify conductive, convective, and radiative heat transfer mechanisms, respectively. Eq. 1.2 is the initial condition; Eq. 1.3 describes heating and cooling at the object's surface boundary; Eqs. 1.4 and 1.5 represent the continuity of temperature and heat flux at the boundaries between inner layers, including defects.

To make use of this theoretical model, given by Eqs. 1.1–1.5, in an autonomous robotic application for categorizing objects we would first solve the model for some physical variable for comparison with measured values from known objects. Class assignment is determined by the computed values' closeness to the measured values. However, this model is nonlinear and rather complicated. As we see, T is a function of many variables, $T(t, \rho, \varepsilon, C, Q, K)$. The problem becomes even more involved with the fact that variables such as conductivity, specific heat, and emissivity may be dependent on time, position, and the object's temperature. Thus, distinct classes of objects heat up and cool at different rates based on their thermal-physical properties. For instance, the surface temperature of low specific heat objects, such as the leaves on hedges, tend to track the availability of solar energy [23]. On the other hand, objects with a high specific heat, such as a birch tree trunk ($\sim$2.4 $kJ \cdot kg^{-1} \cdot {}^\circ C^{-1}$) [22], will tend to heat up more slowly with increasing solar energy and cool more slowly as the amount of solar energy begins to decrease in the late afternoon (around 1600 hrs). Furthermore, for outside objects, windy conditions may influence convective heat transfer.

Simplified model versions of Eqs. 1.1–1.5 are usually used to directly solve for a unique temperature solution using the initial and boundary conditions. There are numerous texts that provide methods to solve the direct problem, two classic texts are [26, 27]. One can also use simplified models to estimate the thermal-physical parameters, which is called the inverse problem. Methods involving inverse problems can be found in [22, 28]. A review of both direct and inverse heat transfer methods is found in [29]. These excellent references provide both analytical and numerical methods to solve simplified heat transfer problems. However, when seeking to generate features from signal data produced by a given object in an unstructured outdoor environment, we must consider the complexities of the real world. Consequently, we must consider the multi-mode heat transfer model and the fact that the thermal-physical variables are dependent upon time, space, and the object's temperature. Rather than attempting to solve the direct or inverse problems mentioned above, we will use the observational model approach to design a pattern classification model that generates thermal-physical features from an objects thermal image. As we will see in Chap. 3, our thermal-physical features are generated from information in the thermal image that encompasses the thermal-physical properties of the object that depend on the diurnal cycle of solar energy.

1.5 Overview of the Book

The primary objective of this book is to design and implement a pattern classification model used by an intelligence algorithm to characterize non-heat generating outdoor objects in thermal scenes for autonomous robotic systems. Our approach to meet this objective is outlined in the model design cycle illustrated in Fig. 1.11.

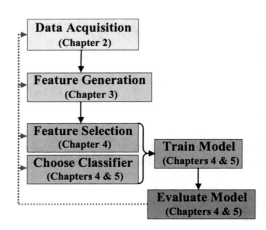

Fig. 1.11 Pattern classification model design cycle.

1.5 Overview of the Book

The chapter that discusses each step is noted in this design cycle flowchart. Since the goal in designing a classification model is to assign unknown objects to classes with minimal classification errors, the results of the evaluation may require repeating certain steps to achieve acceptable performance by the model. In Chap. 2 we will present our robotic thermal imaging system and methodology used to preprocess the thermal signals received by the thermal infrared imaging camera. We will also discuss our procedures to acquire representative data sets for non-heat generating objects to assist in designing and assessing the performance of our classification models. We will present a literature review on feature types and our approach to generating thermal features in Chap. 3. A classification model is defined by at least one classifier and set of features. The performance of a classifier is a function of the feature set. Consequently, the evaluation of classifiers and selection of feature sets are done simultaneously as indicated by the flowchart for the model design cycle. In Chap. 4 we will provide a literature review on approaches to pattern classification and discuss our methodology for selecting thermal features. We will select our most favorable sets of features using the traditional Bayesian, K-Nearest-Neighbor, and Parzen classifiers. In Chap. 5 we will present our Adaptive Bayesian Classification Model that outperforms these traditional classifiers for our application. In Chap. 6 we will offer some conclusions and discuss future research directions.

A possible intelligence algorithm that could be supported by our model is illustrated in Fig. 1.12. The steps with the shaded regions highlight this book's contributions to the intelligence process. A thermal infrared imaging sensor receives thermal energy emitted from an unknown object's surface. The signal received by

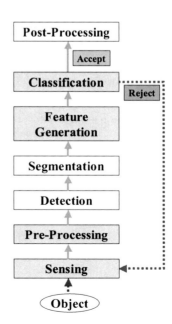

Fig. 1.12 Intelligence algorithm with pattern classification model.

the sensor is preprocessed to minimize the effects of temporal and spatial degradations and dead pixels that would have a negative impact on the bot's ability to generate relevant features from the thermal image and classify unknown objects. The object is detected and segmented in the thermal scene by identifying its thermal contrast with other surfaces in the surrounding environment within the camera's field of view. Features are generated from the segmented object and used by the classification model to assign the unknown object to a specific class with a given degree of confidence represented by the respective posterior probability. If the classification model's decision satisfies specific rules, the class assignment is accepted for post-processing. Otherwise, the class assignment is rejected and the bot is required to capture another image to classify the unknown object. The post-processing step uses the classification model's accepted output to decide on the bot's next required action [*report the object* and/or (*if the object is a hedge, go through the object* or *if the object is a brick wall, go around the object* or *if the object is a trash can, pick up the object*)].

References

[1] Hinders M, Gao W et al (FEB 2007) Sonar Sensor Interpretation and Infrared Image Fusion for Mobile Robotics. In: Kolski S (ed) Mobile Robots: Perception & Navigation. Pro Literatur Verlag, Germany / ARS, Austria

[2] Treptow A, Cielniak G et al (2006) Real-Time People Tracking for Mobile Robots using Thermal Vision. Robotics and Autonomous Systems 54(9):729–739

[3] Michel JD, Nandhakumar N et al (1998) Geometric, Algebraic, and Thermophysical Techniques for Object Recognition in IR Imagery. Computer Vision and Image Understanding 72(1):84–97

[4] Bekey G, Yuh J (2008) The Status of Robotics. IEEE Robotics & Automation Magazine 15(1):80–86

[5] National Research Council. Committee on Army Unmanned Ground Vehicle Technology, National Research Council. Board on Army Science and Technology (2002) Technology Development for Army Unmanned Ground Vehicles. National Academies Press, Washington, D.C.

[6] Xie Z, Li Z et al (2004) Simultaneously-Firing Sonar Ring Based High-Speed Navigation for Nonholonomic Mobile Robots in Unstructured Environments. IEEE Proceedings 2004 International Conference on Intelligent Mechatronics and Automation :650–655

[7] Xin-Ling W (2007) Research and design of intelligent robot control system based on infrared and ultrasonic technology. Proceedings of SPIE, 3rd International Symposium on Advanced Optical Manufacturing and Testing Technologies: Optical Test and Measurement Technology and Equipment:67231H-1-67231H-6

[8] Federici JF, Schulkin B et al (2005) THz Imaging and Sensing for Security Applications – Explosives, Weapons and Drugs. Semiconductor Science and Technology 20(7): S.266–S280

[9] Bhanu B, Pavlidis I (2005) Computer Vision Beyond the Visible Spectrum. Springer, London

[10] Burton M, Benning C (1981) Comparison of imaging infrared detection algorithms. SPIE, Infrared Technology for Target Detection and Classification:26–32

[11] Takahashi S, Nara S (2007) Navigation control for mobile robot based on vision and ultrasonic sensors. Proceedings of SPIE, Optomechatronic:67190I-1-67190I-8

References

[12] Kim P, Park C et al (2007) Obstacle Avoidance of a Mobile Robot using Vision System and Ultrasonic Sensor. In: Anonymous (ed) Advanced Intelligent Computing Theories and Applications with Aspects of Theoretical and Methodological Issues. Springer Berlin / Heidelberg

[13] Tsalatsanis A, Valavanis K et al (2007) Mobile Robot Navigation using Sonar and Range Measurements from Uncalibrated Cameras. Journal of Intelligent & Robotic Systems 48(2):253–284

[14] Bhargave A, Ambrose B et al (2007) Multi-sensor detection and fusion technique. Proceedings of SPIE, Multisensor, Multisource Information Fusion: Architectures, Algorithms, and Applications:657109-1-657109-9

[15] Minor LG, Sklansky J (1981) The Detection and Segmentation of Blobs in Infrared Images. IEEE Transactions on Systems, Man and Cybernetics SMC-11(3):194–201

[16] Lu YJ, Hsu YH et al (1992) Vehicle Classification using Infrared Image-Analysis. Journal of Transportation Engineering-Asce 118(2):223–240

[17] Gonzalez RC, Woods RE et al (2004) Digital Image Processing using MATLAB. Pearson/Prentice Hall, Upper Saddle River, NJ

[18] Odedra S, Prior S et al (2007) Improving the Mobility Performance of Autonomous Unmanned Ground Vehicles by Adding the Ability to 'Sense/Feel' their Local Environment. Virtual Reality :514–522

[19] Lalonde J, Vandapel N et al (2006) Natural Terrain Classification using Three-Dimensional Ladar Data for Ground Robot Mobility. Journal of Field Robotics 23(10):839–861

[20] Manduchi R, Castano A et al (2005) Obstacle Detection and Terrain Classification for Autonomous Off-Road Navigation. Autonomous Robots 18(1):81–102

[21] Ojeda L, Borenstein J et al (2006) Terrain Characterization and Classification with a Mobile Robot. Journal of Field Robotics 23(2):103–122

[22] Maldague XPV (2001) Theory and Practice of Infrared Technology for Nondestructive Testing. Wiley, New York

[23] Holst GC (2000) Common Sense Approach to Thermal Imaging. JCD Pub.; co-published by SPIE Optical Engineering Press, Winter Park, Fla.; Bellingham, Wash.

[24] Traycoff RB (1994) Medical Applications of Infrared Thermography. In: Maldague XPV (ed) Infrared Methodology and Technology. Gordon and Breach Science Publishers, Yverdon, Switzerland ; Langhorne, Pa., U.S.A.

[25] Vavilov VP (1994) Infrared Techniques for Materials Analysis and Nondestructive Testing. In: Maldague XPV (ed) Infrared Methodology and Technology. Gordon and Breach Science Publishers, Yverdon, Switzerland ; Langhorne, Pa., U.S.A.

[26] Carslaw HS, Jaeger JC (1946) Conduction of Heat in Solids. 2nd edn. Clarendon Press; Oxford University Press, Oxford Oxfordshire; New York

[27] Özışık MN (1968) Boundary Value Problems of Heat Conduction. International Textbook Co, Scranton

[28] Özışık MN, Orlande HRB (2000) Inverse Heat Transfer : Fundamentals and Applications. Taylor & Francis, New York

[29] Goldstein RJ, Ibele WE et al (2006) Heat Transfer – A Review of 2003 Literature. International Journal of Heat and Mass Transfer 49(3–4):451–534

2 Data Acquisition

Abstract The first step in our pattern classification model design process is presented – data acquisition. We will first introduce the robotic thermal imaging system, consisting of the hardware and software used to acquire thermal data. We will also discuss the methodology used to preprocess and collect our representative data set. The methodology used in our data acquisition is implemented prior to the feature generation step discussed in the next chapter.

2.1 Introduction

In this chapter, we will present the first step in our pattern classification model design process – data acquisition. We will first introduce our robotic thermal imaging system. This system consists of the hardware and software that is used to acquire the image data. We will also discuss the methodology used to preprocess and collect our representative data set prior to the feature generation step discussed in the next chapter.

2.2 Robotic Thermal Imaging System

2.2.1 Hardware

The hardware for our robotic thermal imaging system is displayed in Fig. 2.1. Figure 2.1 (a) shows the front view of the robot platform. A metal container encloses the thermal camera to ensure that the camera is on a stable platform and protected from the outside environment. The underside of the adjustable lid on the metal container consists of a polished aluminum plate to reflect thermal radiance

26   2 Data Acquisition

emitted from a target to the thermal camera. The polished aluminum plate is a good reflector of thermal radiation due to its low emissivity value (approximately 0.09 for wavelengths of 8–14 μm) [1]. Consequently, the combination of the thermal camera, metal container, and polished aluminum plate act as a periscope. A Futaba remote control module (displayed in the bottom right corner of Fig. 2.1 (a)) is used to navigate the robot platform.

The thermal camera secured at the bottom of the metal container and displayed in Fig. 2.1 (c) is a *Raytheon ControlIR 2000B* long-wave (7–14 micron) infrared thermal imaging video camera with a 50 mm focal length lens. The key specifications of the *Raytheon ControlIR 2000B* include: 320 X 240 pixel resolution, 30 Hz frame rate, 18° x 13.5° field of view (with 50 mm lens), and ferroelectric staring

Fig. 2.1 Robotic thermal imaging system hardware: (a) robot platform front view, (b) robot platform rear view, (c) Raytheon thermal imaging video camera, (d) VideoAdvantage USB video capture device, (e) Samsung tablet PC w/ Powerbank.

focal plane array detector type. As discussed in Chap. 1, Planck's blackbody radiation law tells us that the magnitude of the radiation emitted by an object varies with wavelength for a given temperature. A perfect emitter (or blackbody) with a surface temperature in the interval from $32°$ to $100°F$ radiates a greater magnitude of thermal energy in the wavelength interval of 7–14 microns compared to shorter wavelengths. Therefore, radiation emitted from non-heat generating objects outdoors will peak in the long-wavelength range. In the context of this research, non-heat generating objects are defined as objects that are not a source for their own emission of thermal energy, and so exclude people, animals, vehicles, etc. Consequently, a thermal imaging camera that is sensitive to long-wave thermal radiation is an ideal sensor for our classification application involving non-heat generating objects.

Figure 2.1 (b) displays the rear view of the robot platform. Two metal lockers with hinged doors are stacked behind the "periscope." A *Barnant 90 Digital Thermometer* is attached to the top locker to allow the operator to record the ambient temperature. The bottom locker provides storage for field supplies while the top locker holds a *Samsung Tablet PC* and Powerbank (Fig. 2.1 (e)). *Samsung Tablet PC* has an Intel Celeron 900 MHz processor, 512 MB of RAM, and Microsoft Windows XP Tablet PC Edition operating system. The Powerbank extends the tablet PC's battery life by allowing the operator to continuously capture thermal images for up to approximately 2.5 hours.

The process of capturing a thermal image of a specific target begins with the detectors in the camera's Focal Plane Array (FPA) receiving the thermal radiation emitted from all of the surfaces of objects within the thermal scene. The thermal scene consists of all objects within the camera's field of view, which includes the target of interest and objects in the foreground. In the context of this research, we define foreground as the region in the scene consisting of objects behind the target of interest and within the thermal camera's field of view. Background is defined as the region either in front or to the side of the target consisting of thermal sources that emit thermal energy onto the target's surface. The source emitting this thermal energy may or may not be in the camera's field of view. The thermal radiation received by the FPA is converted to an analog signal with a 320X240 pixel resolution. This analog signal is transmitted from the camera through a harness cable assembly to a *Voyetra Turtle Beach Video Advantage USB Video Capture* device (see Fig. 2.1 (d)) that is attached to the *Samsung Tablet PC*. The *Voyetra Turtle Beach Video Advantage USB Video Capture* device converts the composite analog signal from the camera to a digital signal. The tablet PC receives the digital signal and a thermal image is displayed on the screen using the *VideoAdvantage* software that is installed on the tablet PC, discussed below. A camera control cable also connects the camera to the *Samsung Tablet PC*. The *Control IR Manager* software installed on the tablet PC, discussed in the following section, uses this cable to make modifications to the camera's memory. During thermal image capturing sessions, the door on the top locker is closed to prevent glare on the tablet PC's display screen caused by the sun. With the door shut, the operator views the thermal image on the tablet PC's display screen through the black eyepiece and captures thermal images with the *VideoAdvantage* software using the mouse, both located on the top locker (see Fig. 2.1 (b)).

2.2.2 Signal Preprocessing

In this section, we will discuss the software used to capture and preprocess a thermal image of an object prior to generating features. The significance of preprocessing a thermal image is evident from the thermal image in Fig. 2.2. The quality of this thermal image is affected by temporal and spatial signal degradations and dead pixels. If the magnitude of these typical degradation processes is not minimized, they will have a negative impact on our ability to generate relevant features from the thermal image and characterize unknown objects.

Fig. 2.2 Thermal image prior to preprocessing.

2.2.2.1 Signal Degradations

Signal degradations consist of temporal and spatial signal degradations and dead pixels. Temporal signal degradations consist of a temporal fluctuation in the signal at a low frequency (drift), mechanical vibrations due to the movement of camera system relative to the target (jitter), and noise (electronic, optical, and structural) [2, 3]. The spatial signal degradations are displayed as the fine horizontal and vertical lines overlaid on the thermal image in Fig. 2.2. These spatial signal degradations are due to non-uniformities in the responsivity of the detectors in the FPA [2, 3]. We also see dead pixels (white specks throughout the image) resulting from a defect in the instrumentation caused by events such as heat deterioration or a high incidence of static electricity on a detector [4].

Control IR Manager is software used to control the functionality of the *Raytheon ControlIR 2000B* infrared thermal imaging video camera. The software is used to make modifications to the camera's memory that will preprocess the thermal images and minimize the effects of the degradation processes. We will discuss the key software features used to preprocess our thermal images. Figure 2.3 displays the main menu of the *Control IR Manager* software. The polarity switch in the upper left corner is set to White Hot, resulting in objects with apparent high temperature (hot) surfaces, relative to other objects in the camera's field of view, to yield gray-scale values of 255 (white) in the thermal image. On the other hand, objects that have an apparent low temperature (cold) surface, relative to other objects in the camera's field of view, will yield gray-scale values 0 (black) in the thermal image. Consequently, the thermal radiance emitted from surfaces of objects in the entire scene could result in various gray-scale values in the interval [0, 255]. This characteristic of the thermal camera will lead us to an important discussion on AC coupling and the AGC circuit that we will cover shortly.

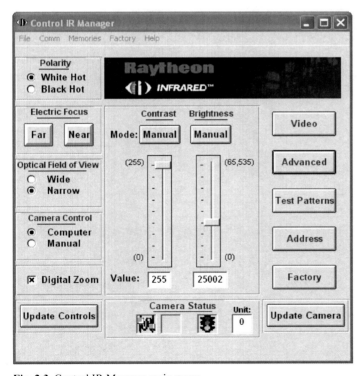

Fig. 2.3 Control IR Manager main menu.

30   2 Data Acquisition

We will now discuss how to make adjustments in the software to suppress the temporal and spatial signal degradations and dead pixels. By selecting the Video icon from the *Control IR Manager* software's main menu (Fig. 2.3) we get the Video Settings menu (Fig. 2.4). By enabling Frame Integration with 16 frames, we can reduce the effects of temporal signal degradations by taking a frame-to-frame average of the scene over 16 frames (the *Raytheon ControlIR 2000B* has a frame rate of 30 Hz). Moving back to the main menu (Fig. 2.3) and selecting the Advanced icon, we go to the Advanced Video Settings menu (Fig. 2.5). By enabling Normalization Correction in the Normalization Options menu, we are able to treat the spatial signal degradations due to the non-uniformity of the detectors in the FPA. Our system uses a single-point non-uniformity correction method that normalizes (makes equal) the outputs for the individual detectors over a uniform thermal scene [5]. In single-point correction, the average of multiple images of a uniform thermal scene (single thermal input intensity or temperature reference) is subtracted from live video to remove the non-uniformity [2, 3]. Also within the Normalization Options menu (Fig. 2.5), we can enable Pixel Substitution to store locations of dead pixels in the FPA and substitute the dead pixels with the mean value of horizontally adjacent good pixels. After suppressing the temporal and spatial signal degradations and dead pixels found in

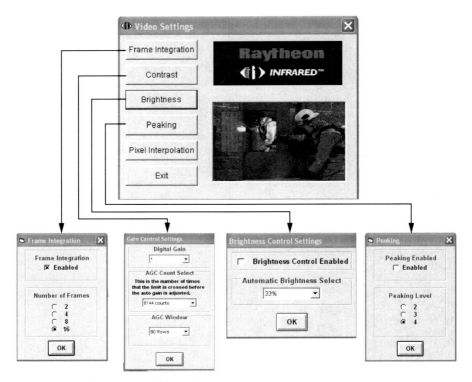

Fig. 2.4 Control IR Manager video settings.

Fig. 2.2, we obtain the resulting thermal image in Fig. 2.6. Table 2.1 presents theprocedure to normalize the camera and store the reference in the camera's memory to perform non-uniformity correction on subsequent thermal image frames.

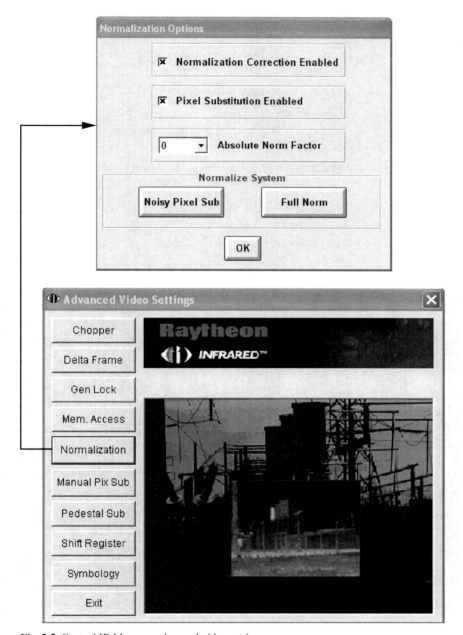

Fig. 2.5 Control IR Manager advanced video settings.

Table 2.1 Procedure to normalize the camera and store the reference in the camera's memory to perform non-uniformity correction on subsequent thermal image frames [5].

(1) Setup the *Raytheon ControlIR 2000B* approximately 2.5 meters from a smooth, non-shiny, surface with a low thermal reflectivity (i.e., high emissivity), such as plywood with black spray paint on the surface. This uniform surface must take up the entire scene in the camera's field of view.
(2) Select the Factory icon (Fig. 2.3) and disable Norm Threshold in the Factory Options menu.
(3) At the main menu (Fig. 2.3), disable Digital Zoom.
(4) Select the Advanced icon (Fig. 2.3) and the Normalization icon in Advanced Video Settings (Fig. 2.5). Enable Normalization Correction and Pixel Substitution in the Normalization Options menu.
(5) In the Normalization Options menu (Fig. 2.5), select the Full Norm icon under Normalize System. Run Full Norm for at least 5 minutes and then select Stop.
(6) At the main menu (Fig. 2.3), enable Digital Zoom.
(7) Again, in the Normalization Options menu (Fig. 2.5), select the Full Norm icon under Normalize System. Run Full Norm for at least 5 minutes and then select Stop.
(8) Select the Factory icon (Fig. 2.3) and enable Norm Threshold in the Factory Options menu.

Fig. 2.6 Thermal image with preprocessing on temporal/spatial signal degradations and dead pixels. AGC is enabled.

2.2.2.2 AC Coupling

As mentioned earlier, the polarity for the *Raytheon ControlIR 2000B* was set so the thermal radiance of the surfaces of objects in the entire scene could map to various gray-scale values in the interval [0, 255] where the extremes 0 (black) and 255 (white) imply apparent cold and hot surfaces, respectively. Furthermore, we mentioned that the gray-scale values of an object in a thermal image are assigned relative to other objects in the camera's field of view. This is a characteristic of thermal cameras with FPAs that is known as AC coupling. AC coupling is integrated into the *Raytheon ControlIR 2000B* so that small variations of the surface radiance of objects in a scene can be amplified [1]. Also, a thermal image is AC coupled horizontally along the rows in the image array. A consequence of AC coupling is that a specific target in a scene with a constant thermal radiance could be assigned a large or small gray-scale value depending on the other surfaces in the surroundings within the camera's field of view. Furthermore, a target can only be seen in a thermal image when a thermal contrast exists between the target and other objects in the camera's field of view. Consequently, useful feature values to distinguish objects can only be generated when a thermal contrast exists in the thermal scene. Of course, this makes the objective to classify non-heat generating objects even more challenging since these objects depend highly on prior solar energy absorption in order to emit thermal radiation.

As a result of AC coupling, a target is not radiometrically correct (i.e., the gray-level value is not a linear function of the apparent surface temperature). Figure 2.7 shows an example of AC coupling similar to one illustrated in [1]. Figure 2.7 (a) simulates a scene with uniform thermal physical surface properties (i.e., emissivity, specific heat, etc.) but with different temperature regions. Figure 2.7 (b) displays the resulting thermal image of this scene after AC cou-

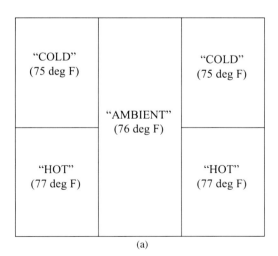

(a)

Fig. 2.7 (a) AC coupling. Scene with different temperature regions.

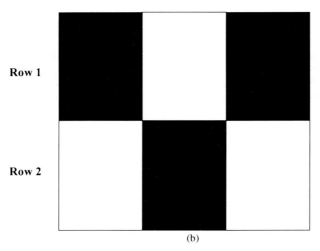

Fig. 2.7 (b) AC coupling. Gray-level shades of regions in thermal image.

pling. As we see, the ambient region maintains a constant temperature of 76°F. However, with AC coupling applied horizontally along the rows in the image, the regions in each row are assigned gray-levels relative to other objects in the same row. As a result, the upper half of the ambient region appears "hot" in Row 1 and the lower half of the ambient region appears "cold" in Row 2.

2.2.2.3 Automatic Gain Control

The effects of AC coupling alone will not hinder our ability to generate features to distinguish objects. However, problems do arise when the amplifications of the gray-level values for an object at a constant temperature become extreme. This issue exists when the *Raytheon ControlIR 2000B*'s automatic gain control (AGC) circuit is enabled. The AGC is an image enhancer that is designed to afford the operator with comfortable image viewing. The AGC automatically adjusts the gain (and offset) to ensure the signals are within the camera's dynamic range to minimize saturation of objects in the scene [2]. As a result of the AGC, the thermal image of a bright object may be darker and dark object may be brighter. Thus, the AGC amplifies the effects of AC coupling. Similar to AC coupling, the AGC results in gray-level values assigned to objects relative to other objects within a given window. The effects of the AGC circuit are illustrated in Fig. 2.6. Even though the actual surface of the pole is approximately uniform in thermal properties (to include temperature), its thermal image displays an apparent temperature difference between the bottom portion of the pole (with the building in the foreground) and top portion of the pole (with the sky in the foreground).

To investigate the effects of the AGC circuit further, we analyzed variations in gray-level values of a cardboard tube with a constant surface temperature adjacent to

2.2 Robotic Thermal Imaging System

a cardboard tube that is heated to a given temperature and allowed to cool. The cardboard tubes were secured in a thermally insulated box with an opening in the front and a thermal insulator separating the tube on the left (at a constant temperature) from the tube on the right (that was heated). The experiment was conducted in a controlled environment with a constant surrounding radiance and ambient temperature of approximately $59.3°F$. The left cardboard tube with a constant surface temperature was maintained at approximately $86.5°F$. The surface of the right cardboard tube was heated to $110.8°F$ and allowed to cool to $65.8°F$. Ten images of the scene consisting of the two tubes were captured at increments of approximately $5°F$ based on the right cardboard tube that was cooling. The mean gray-level values were recorded on the same segments of the two tubes for each image captured. Figure 2.8 illustrates the experimental results with the AGC enabled. Figure 2.8 (a) and 2.8 (b) display the first (right tube at $110.8°F$) and tenth (right tube at $65.8°F$) images captured, respectively. By comparing Fig. 2.8 (a) and 2.8 (b), we see that the tube on the left (maintained at a constant temperature) varies in gray-levels due to the AGC. Figure 2.8 (c) displays the variations of the gray-levels for the constant and heated tubes as a function of temperature. With the AGC enabled, the constant tube has a standard deviation of 13.84 and range of 44.98 in the gray-levels. Consequently, these extreme variations in gray-level values for the constant tube would hinder our ability to generate relevant features to distinguish objects. Fortunately, we can make modifications to the *Raytheon ControlIR 2000B*'s memory, using the *Control IR Manager* software, to disable the AGC by following the procedure presented in Table 2.2.

We conducted another experiment under the same conditions as described above with the cardboard tubes, with the exception that the AGC was disabled. Once again, the left cardboard tube with a constant surface temperature was maintained at approximately $86.5°F$. The surface of the right cardboard tube was heated to $110.4°F$ and allowed to cool to $65.8°F$. Ten images of the scene consisting of the two tubes were captured at increments of approximately $5°F$ based on the right cardboard tube that was cooling. Figure 2.9 illustrates the experimental results with the AGC disabled. Figure 2.9 (a) and 2.9 (b) display the first (right tube at $110.4°F$) and tenth (right tube at $65.8°F$) images captured, respectively. By comparing Fig. 2.9 (a) and 2.9 (b), we see that the tube on the left (maintained at a constant temperature) appears to have minimal variation in gray-levels when the AGC is disabled. Figure 2.9 (c)

Table 2.2 Procedure to disable AGC by making modifications in the *Raytheon ControlIR 2000B*'s memory using the *Control IR Manager* software [5].

(1) In the Video Settings menu (Fig. 2.4), select the Contrast icon to display the Gain Control Settings. Set the Digital Gain equal to 1, AGC Count Select to 6144 counts, and AGC Window to 80 Rows.

(2) In the Video Settings menu (Fig. 2.4), select the Brightness icon to display Brightness Control Settings. Disable the Brightness Control.

(3) In the Control IR Manager main menu (Fig. 2.3), set the Contrast Mode to Manual with a Value of 255 and Brightness Mode to Manual with a Value of 25002.

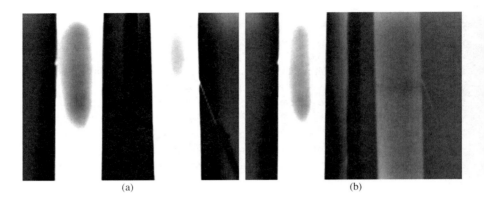

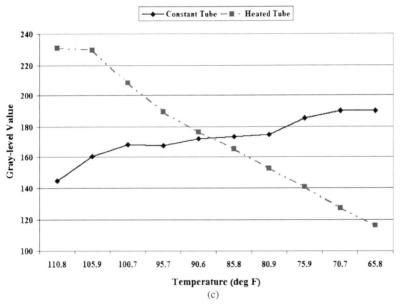

Fig. 2.8 Enabled AGC experiment with cardboard tubes (left tube at constant temperature of ~86.5 deg F and right tube heated to 110.8 deg F and allowed to cool to 65.8 deg F). (a) Image of tubes with right (heated) tube at 110.8 deg F, (b) Image of tubes with right (heated) tube at 65.8 deg F, (c) Variations of gray-levels of constant and heated tubes as a function of temperature.

displays the variations of the gray-levels for the constant and heated tubes as a function of temperature. With the AGC disabled, the constant tube has a standard deviation of 2.16 and range of 7.22 in the gray-levels. Thus, by disabling the AGC, the variations in the gray-level values for the constant tube are only due to AC coupling. Furthermore, by disabling the AGC, the thermal image of the pole displayed in Fig. 2.6 now provides acceptable results for the variation of gray-levels as displayed

2.2 Robotic Thermal Imaging System   37

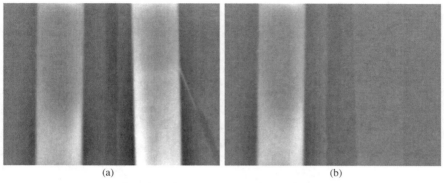

Gray-level Variations with AGC Disabled

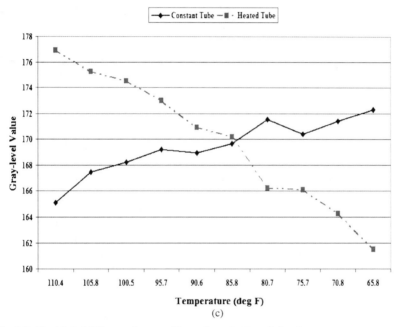

Fig. 2.9 Disabled AGC experiment with cardboard tubes (left tube at constant temperature of ~ 86.5 deg F and right tube heated to 110.4 deg F and allowed to cool to 65.8 deg F). (a) Image of tubes with right (heated) tube at 110.4 deg F, (b) Image of tubes with right (heated) tube at 65.8 deg F, (c) Variations of gray-levels of constant and heated tubes as a function of temperature.

in Fig. 2.10. Therefore, with the AGC disabled we can now generate relevant features from the thermal images of objects that will assist us in classifying the objects.

At this point it is appropriate to mention the halo effect around the bottom portion of the pole in Fig. 2.6. The halo effect is common with ferroelectric FPAs where accurate imagery is assisted by a mechanical chopper wheel within the camera. As discussed in [6], capturing a thermal image of a target is a cyclic process.

38   2 Data Acquisition

Fig. 2.10 Thermal image with preprocessing on temporal/spatial signal degradations and dead pixels. AGC is disabled.

Suppose the target is emitting more thermal radiation than any other neighboring object in the scene (either directly adjacent or behind the target). To capture a thermal image, the target first emits radiation onto the back of the chopper wheel and the FPA obtains a charge reading from the wheel. Next, the FPA obtains a charge directly from the actual target emitting the thermal radiation. Lastly, the system electronically subtracts the charges with and without the chopper wheel to produce the thermal image. However, the thermal radiation from the hot target that leaks through the camera's chopper wheel is unfocused, leaving a larger radiation imprint on the FPA than that of the actual target. When the system subtracts the charges with and without the chopper wheel, a halo is created around the "hot" target in the image that is darker than the "cold" foreground. As we will see in Chap. 3, a "cold" target and "hot" foreground will result in a halo around the "cold" target that is a lighter shade than the "hot" foreground. Fortunately, the halo effect will not interfere with our ability to generate thermal features for classifying objects. As a matter of fact, we will discuss in Chap. 6 how we may be able to use the halo effect to facilitate the segmentation of targets [7].

2.2.2.4 Filters

One of our goals in preprocessing is to suppress degradations in the signal without losing information that would assist in classifying objects in the scene. Consequently, we avoid filters that would lead to loss of relevant information used to distin-

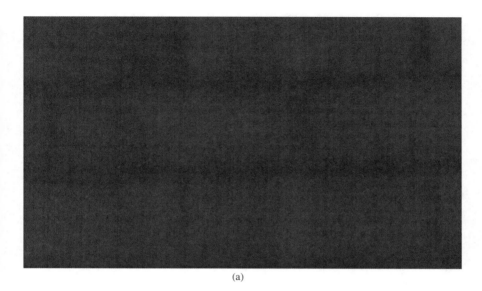

(a)

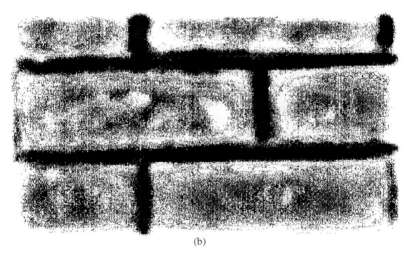

(b)

Fig. 2.11 Thermal image of segment of brick wall: (a) without high pass filter, (b) with high pass filter.

guish object classes. For example, in the Video Settings menu (Fig. 2.4) of the *Control IR Manager* software we disable Peaking since this functionality performs a high-pass filter on the thermal image. Figures 2.11 (a) and (b) display thermal images of the same segment of a brick wall without and with a high pass filter, respectively. As we see in Fig. 2.11 (b), applying a high pass filter results in a loss of information that could be used by thermal features to classify objects.

40   2 Data Acquisition

(a)

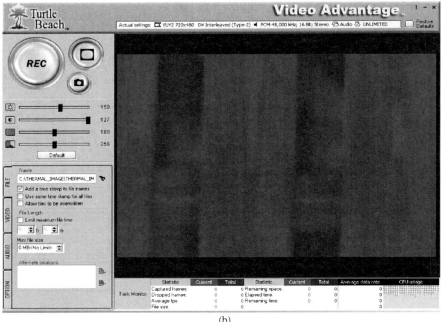

(b)

Fig. 2.12 (a) Robotic thermal imaging system capturing an image of a wood fence. (b) Thermal image of the wood fence displayed with *VideoAdvantage* software.

2.2.2.5 Capturing Thermal Imagery

After the analog signal from *Raytheon ControlIR 2000B* is converted to a digital signal by the *Voyetra Turtle Beach Video Advantage USB Video Capture,* the *Samsung Tablet PC* receives the digital signal and a thermal image is displayed on the screen using the *VideoAdvantage* software. Figure 2.12 illustrates a scenario with the robotic thermal imaging system capturing an image of a segment of a wood fence. The *VideoAdvantage* software displays live video and is capable of capturing continuous or still frames. Our current research will focus on classifying non-heat generating objects in thermal images using still frames. However, we intend to extend our research to classify objects using continuous frames as discussed in Chap. 6. The final preprocessing step before the feature generation phase is to convert the RGB (red, green, blue) image captured by the *VideoAdvantage* software to a gray-scale image using MATLAB.

2.3 Data Collection

We now present the methodology used to collect the data needed to train and evaluate our pattern classification model. We assume that the robot already makes use of algorithms to detect and segment a specific target, analogous to those discussed in Chap. 1. In Chap. 6, we will discuss possible techniques for automated detection and segmentation of objects that we intend to integrate into our future research. Consequently, in the current research we will manually segment our targets.

Thermal imagery was captured on a variety of non-heat generating outdoor objects during a nine-month period, at various times throughout the days and at various illumination/temperature conditions. The ambient temperature (in degrees Fahrenheit) was recorded during each session. The images were captured using a Raytheon *ControlIR* 2000B long-wave (7–14 micron) thermal infrared imaging video camera with a 50 mm focal length lens at a distance of 2.4 meters from the given objects. The analog signals with a 320 X 240 pixel resolution were converted to digital signals using a *Voyetra Turtle Beach Video Advantage USB Video Capture* device attached to a *Samsung Tablet PC*, all mounted on board a mobile robotic platform displayed in Fig. 2.1. The resulting digital frames were preprocessed as discussed in Sect. 2.2.

The image data was divided into two categories: extended objects and compact objects. The extended objects consist of objects that extend laterally beyond the thermal camera's lateral field of view. Our classes of extended objects consist of brick walls, hedges, wood picket fences, and wood walls. The compact objects consist of objects that are completely within the thermal camera's lateral field of view. Our classes of compact objects consist of steel poles and trees. The image data collected was partitioned into three mutually exclusive sets: training data, test data, and blind data. The training data was used to design our pattern classification model. The performance of the model was assessed using the test and blind data sets. Since the test set was used as a validation set to tune the pattern classification model, it was part of the training process and not being used to provide an independent error estimate. Therefore, the blind data set was used for our independent performance evaluation of the pattern classification model.

Our objective is to design a pattern classification model that displays exceptional performance in classifying unknown non-heat generating objects in an outdoor environment. To satisfy this objective, the data that we collect must completely and accurately represent the real world problem by consisting of all the meaningful variations of field data instances that the system is likely to encounter. Thus, our representative data was collected under diverse environments (climates), temperatures, solar energy conditions, and viewing angles.

Figures 2.13 and 2.14 display the visible and typical thermal images of extended and compact objects, respectively, used for our training data that was collected from 15 March to 22 June 2007 about The College of William & Mary campus. The strips of black electrical tape shown in the visible images and displaying a high thermal radiance in some of the thermal images are used as a reference emitter for generating the thermal features that we will discuss in Chap. 3. During each of the 55 sessions, the thermal images were captured on each object from two different viewing angles: normal incidence and 45 degrees from incidence. Table 2.3 and Fig. 2.15 present the frequencies of the object classes and ambient temperature distribution for the training data, respectively.

The thermal images used for the test data consisted of the same objects used in the training data (Figs. 2.13 and 2.14). The thermal images were captured at the same viewing angles as the training data. However, the test data was collected over nine sessions from 25 June to 3 July 2007. Table 2.3 and Fig. 2.15 present the frequencies of the object classes and ambient temperature distribution for the test data, respectively.

The blind data set was collected over 14 sessions from 6 July to 5 November 2007. The thermal images used for the blind data set consisted of the same classes and were captured at the same viewing angles as the training data but were not the same objects. In addition to some blind data being collected on The College of William & Mary campus, data was also collected throughout York County, Virginia, in a village and on a farm outside Buffalo, New York, and on mountainous terrain in Eleanor, West Virginia. Table 2.4 presents the frequencies of the objects in the blind data set as well as the locations that the data was collected. Figure 2.15 displays the ambient tem-

Table 2.3 Distribution of training and test data collected from 15 March to 3 July 2007.

DATA TYPE	OBJECT CLASSES						
	Extended Objects				Compact Objects		
	Brick Wall	Hedges	Picket Fence (Wood)	Wood Wall	Steel Pole	Tree	Total
Training	105	107	107	105	318	318	1060
Test	18	16	16	18	48	52	168

2.3 Data Collection 43

(a) (b)

(c) (d)

Fig. 2.13 Visible and thermal images of extended objects from the training data set. (a) brick wall, (b) hedges, (c) wood picket fence, and (d) wood wall.

Fig. 2.14 Visible and thermal images of compact objects from the training data set. Steel poles: (a) brown painted surface, (b) green painted surface, (c) octagon shape w/ aged brown painted surface. Tree: (d) basswood tree, (e) birch tree, (f) cedar tree.

perature distribution of the blind data set. Additionally, to evaluate the classification model's response when confronted with other blind objects, to include objects outside the classes in the training data set, we included data consisting of a brick wall with moss on the surface, concrete wall, bush, gravel pile, steel picket fence, wood bench, wood wall of a storage shed, square steel pole, aluminum pole for a dryer vent, concrete pole, knotty tree, telephone pole, 4 × 4 wood pole, and pumpkin.

Table 2.4 Distribution of blind data collected from 6 July to 5 November 2007.

	OBJECT CLASSES						
	Extended Objects				Compact Objects		
LOCATION	Brick Wall	Hedges	Picket Fence (Wood)	Wood Wall Fence	Steel Pole	Tree	Total
New York	3	4	10	3	1	7	28
William & Mary	16	10	9	14	2		51
West Virginia						1	1
York County	4	9	4	6	17	12	52
Total	23	23	23	23	20	20	132

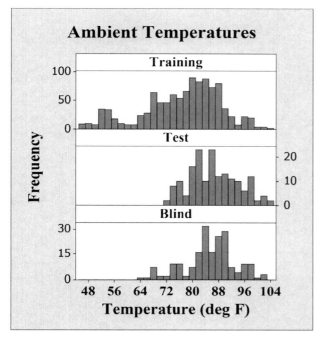

Fig. 2.15 Ambient temperature distributions for training, test, and blind data collected from 15 March to 5 November 2007.

2.4 Summary

In this chapter, we discussed the first step in our pattern classification model design process – data acquisition. We introduced our robotic thermal imaging system consisting of the hardware and software used to acquire thermal data. We also discussed the methodology used to preprocess and collect our representative data set. The methodologies used in our data acquisition are implemented prior to the feature generation step discussed in the next chapter.

References

[1] Holst GC (2000) Common Sense Approach to Thermal Imaging. JCD Pub.; co-published by SPIE Optical Engineering Press, Winter Park, Fla.; Bellingham, WA
[2] Holst GC (1998) Testing and Evaluation of Infrared Imaging Systems. 2nd edn. JCD Pub.; SPIE Optical Engineering Press, Winter Park, FL; Bellingham, WA
[3] Maldague XPV (2001) Theory and Practice of Infrared Technology for Nondestructive Testing. Wiley, New York

[4] Ginesu G, Giusto DD et al (2004) Detection of Foreign Bodies in Food by Thermal Image Processing. IEEE Transactions on Industrial Electronics 51(2):480–490
[5] Private Conversation with Field Application Engineer, L-3 Communications Infrared Products, 27 January 2007.
[6] Pandya N, Van Anda J (2004) Across the Spectrum. SPIE's Oemagazine :28–31
[7] Goubet E, Katz J et al (May 2006) Pedestrian tracking using thermal infrared imaging. Proceedings of SPIE, Infrared Technology and Applications XXXII:62062C-1 - 62062C-12

3 Thermal Feature Generation

Abstract This chapter introduces the thermal features used to classify non-heat generating objects. Examples are provided to illustrate the value of these features in distinguishing non-heat generating objects. By generating feature values from the thermal images of non-heat generating objects, we will see how interpreting the effects of the outdoor environment and thermal properties of objects on their feature values is a subtle process. We will also introduce a curvature algorithm that can be used distinguish compact objects from extended objects.

3.1 Introduction

In Chap. 2 we discussed the procedures for acquiring our thermal images. We will now present the second step in our pattern classification model design process – thermal feature generation. These features are unique representations of a non-heat generating object that are derived from the given object's thermal image. In the context of this research, non-heat generating objects are defined as objects that are not a source for their own emission of thermal energy, and so exclude people, animals, vehicles, etc. Our goal is to generate thermal features that not only assist in distinguishing one object class from another but also have a physical interpretation. We will discuss three types of features – meteorological, micro, and macro. We will also present a curvature algorithm that will allow us to distinguish compact objects from extended objects. Compact objects consist of objects that are completely within the thermal camera's lateral field of view, such as steel poles and trees. Extended objects consist of objects that extend laterally beyond the thermal camera's lateral field of view, such as brick walls, hedges, picket fences, and wood walls. By generating feature values from the thermal images of non-heat generating objects, we will witness how trying to interpret the effects of the outdoor environment and thermal properties of objects on these feature values is a subtle process. In the next chapter, we will evaluate the features' classification performance and select the most favorable set of features.

3.2 "Ugly Duckling" Features

Our thermal-physical features are generated from an object's signal data received by a *Raytheon ControlIR 2000B* long-wave infrared thermal imaging video camera. Through the process of feature generation, the underlying physics of the information in the thermal signal produced by a given object is analyzed to generate unique representations of the object. These features are used to distinguish one object class from another. Ideally, features are chosen that have minimal variation with changes in the viewing angle and/or distance between the object and sensor, temperature, and visibility. Since our objects do not generate their own heat energy, their thermal signature depends on their thermal properties and external heat sources such as the sun and other objects in the surrounding environment. As a result, the amount of thermal radiation emitted from our objects during conditions of limited visibility will depend on the time history of radiation received from external sources. Consequently, the complexity of our application increases due to the variation in thermal radiance of objects in the scene primarily caused by the diurnal cycle of solar energy.

Thermal feature generation is a crucial step in our quest to design a pattern classification model that will allow us to classify non-heat generating objects in an outdoor environment. As we will see in Chap. 4, the performance of a classifier is a function of the feature set. According to the Ugly Duckling Theorem [1], there is no problem independent, universal, or "optimal" set of features. If a set of features appears to perform better in a classification model than another, it is a result of its fit to the particular pattern classification application. In our case, not only do we desire a set of features that maintain their discriminating information, we also seek features that retain their physical interpretation.

There are many choices for the type of features to use in a classification model. Reviews of the various types of features are found in [2, 3, 4, 5, 6, 7]. Two popular types of features used in pattern recognition are moment invariants and Fourier-Mellin descriptors. Moment invariants are geometric features that were first introduced to the pattern recognition community by Hu [8]. Hu's seven famous moments were derived from the normalized central moments of an object's image [3]. Since then, various improvements have been made to Hu's work. Mistakes in Hu's theory were corrected by Reiss [9]. Flusser [10] showed that Hu's system of seven moments is not independent, implying redundancy in the set of features. Considerable research has focused on moments as geometric descriptors that are invariant with respect to translation, rotation, scaling, illumination and blurring of an object in an image [11, 12, 13, 14]. However, moments have a tendency to be sensitive to noise [15]. Another set of features that permit objects in images to be classified according to their shapes are the Fourier-Mellin descriptors, introduced by Casasent and Psaltis [16, 17]. Fourier-Mellin descriptors are generated from the frequency domain of an object's image and used for invariant pattern recognition [18, 19]. The Fourier-Mellin descriptors are also related to Hu's moment invariants [20]. Neither moments nor Fourier-Mellin descriptors are a desirable choice for our features since they lack physical meaning for cases above the third order.

The majority of the classification research involving thermal imagery has involved generating features based on the radiance emitted from heat generating objects or non-heat generating objects that require a thermal excitation in a controlled environment. Heat generating objects could include people, ground vehicles, or marine vehicles. The classification problems in the literature involving people usually involve identification and tracking [21] and facial recognition [22]. Research involving the classification of ground vehicles is found in [23, 24]. Fang and Wu [25] approached armored vehicle classification by generating geometric features, based on Hu's seven moment invariants, from the thermal images of English letters used to represent the contours and wheels of armored vehicles. The features were entered into a neural network where final recognition of a letter was achieved through repeated computation and learning. Classification of ships by comparing their silhouettes against a library of templates is discussed in [26]. Common to these referenced applications is that classification is based purely on geometric features, rather than thermal-physical features generated from the target's surface.

There have been only a few research studies found in the literature involving thermal-physical features generated from a target's surface for classification applications. Nandhakumar and Aggarwal [27] generated features based on estimated values of surface heat fluxes to interpret surfaces in an outdoor scene. Surface temperatures were estimated from a thermal image by assuming that all objects in the scene have an emissivity of approximately 0.9. A visual image of the same scene was used to estimate surface absorptivity and relative orientation of the viewed surface. These estimations were used together to estimate the heat fluxes at the surfaces in the scene. The assumption of a relatively constant emissivity was continued in follow-on research to generate thermo-physical and geometric invariant features from thermal images to classify ground vehicles [28]. Geometric features based on lines and conics were generated from a given region in a thermal image to hypothesize the type of ground vehicle and its pose. Thermal-physical features are formed from both temperature estimates generated from the thermal image and material properties associated with the hypothesized vehicle type and pose. The resulting thermal-physical features are compared with a model prototype based on features expected from the hypothesized vehicle to assess the hypothesized vehicle class. Bharadwaj and Carin [29] generate temperature features estimated from the thermal radiance emitted from various regions on ground vehicles. Vehicles are classified based on the correlation between the feature vectors generated from the different regions on a vehicle and a given template. Maadi and Maldague [30] generate features based on temperature estimations and geometries to classify people and ground vehicles. A multisensor data fusion system using infrared cameras, visual (CCD) cameras, and laser radar sensors for classifying ground vehicles is described in [31]. The features used by this system include geometric attributes, temperature estimations, and colors generated from the target.

There are also many machine vision industrial applications that rely on thermal features generated from the surface of objects to monitor quality control [32, 33, 34]. These applications normally involve feature generation in a controlled indoors environment using a thermal excitation to monitor packaging standards and detect anomalies in products. For instance, in the food industry thermal features could be generated to monitor the seals on food containers [35] or detect anomalies in food [36].

The feature generating techniques in the previous research discussed above are not an appropriate choice for our application. Classification of objects in thermal imagery has mainly involved geometric features, rather than thermal-physical features generated from the target's surface. Consequently, classification of objects has traditionally involved detecting and segmenting thermal "blobs" in the image and generating shape features that are compared to those in database or library of templates. This limitation was mainly a result of the state of the art available in thermal image based systems. Thermal imaging systems did not have the resolution to obtain detailed information about an object's surface. However, our object classes do have a noticeable distinction when comparing their surfaces in a thermal image. Thus, it appears that appropriate features for our application will consider information about the objects' surfaces found in the thermal image.

The previous research that did involve the generation of thermal features from an object's surface required the visible spectrum and/or included temperature estimates. However, to classify non-heat generating objects during conditions of limited visibility, we should not generate features that rely on the visible spectrum. Moreover, thermal cameras do not read temperature on an objects surface directly. To generate an estimated temperature feature from thermal imagery, one must enter a measured or assumed emissivity of the target's surface [37]. Emissivity is a surface property that provides a measure of an object's ability to emit thermal energy. Furthermore, emissivity is a function of the type of material, viewing angle, and the object's surface quality, shape, and temperature [38, 39, 40]. The level of radiance presented by an object's surface in a thermal image depends on the object's emissivity. Consequently, we should not assume an emissivity for an unknown object that we desire to classify. The remote sensing community has successfully used emissivity as a feature to assist in discriminating between vegetation and bare soil [41]. Therefore, an appropriate choice for a feature derived from the thermal image of a non-heat generating object in an outdoor environment seems to be emissivity, not an apparent surface temperature.

Besides the emissivity feature used in remote sensing, we have not identified any other previous research involving the generation of surface features from the thermal imagery of a non-heat generating object in an outdoor environment. However, in the visible spectrum, discriminating information about an object's surface has been obtained using texture features. Weszka, Dyer, and Rosenfeld [42] provide an informative study that compares visual texture features for terrain classification in the field of remote sensing. They concluded that texture features based on first-order and second-order statistics displayed good terrain classification results. The term texture is difficult to define and takes on many definitions in the literature. Furthermore, the concept of texture has been traditionally motivated by human's visual perception of material surfaces [5, 43]. We adopt the definition of texture as a feature dependent on the spatial variation in pixel intensities (gray-level values) [5]. Using this definition of texture allows us to denote an object's variation in surface radiance as the spatial variation in pixel intensities (gray-level values) observed in the object's thermal image. Since our object classes do have a noticeable distinction when comparing their surfaces in a thermal image, first- and second-order texture features seem to be appropriate for our application.

Since we are working with thermal images of non-heat generating objects, the radiance of the objects not only depends on the diurnal cycle of solar energy but also is a function of the object's thermo-physical properties. Consequently, features based on emissivity and texture seem appropriate since they are generated from information in the thermal image that encompass the thermo-physical properties of the object that depend on the diurnal cycle of solar energy. In this research, the generation of these features from segmented objects in thermal images are computed offline in MATLAB.

The remainder of this chapter will proceed as follows. In Sect. 3.3, we will discuss the characteristics of our thermal gray-scale image used for generating features. Section 3.4 will present our meteorological features consisting of the ambient temperature and a rate of change in the ambient temperature. In Sect. 3.5, we will discuss our micro features based on the emissivity of our target's surface. Section 3.6 will present our macro features based on first- and second-order texture features. We will provide an application involving our meteorological, micro, and macro features in Sect. 3.7. Section 3.8 will present a curvature algorithm that will allow us to distinguish compact objects from extended objects. Section 3.9 will provide a summary of the chapter.

3.3 Thermal Image Representation

In this section we will define how our thermal gray-scale (or gray-level) images are represented throughout our research. Figure 3.1 (a) displays our robotic imaging system capturing a thermal image of a fence segment denoted with the

52   3 Thermal Feature Generation

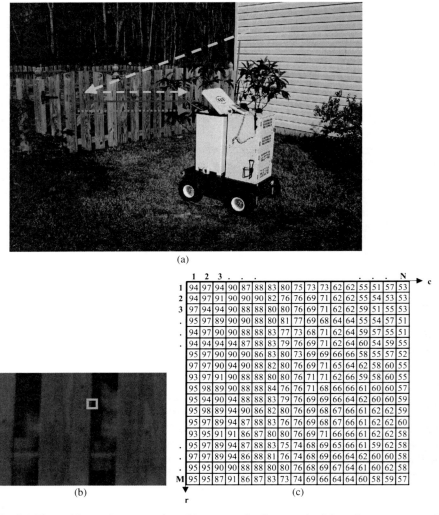

Fig. 3.1 Thermal Image Representation: (a) sources of radiance emitted from fence segment and received by the camera, (b) thermal image of fence segment, (c) data array of gray-level intensities from segment of thermal image.

rectangular solid border. Figure 3.1 (b) presents the resulting thermal image of the fence segment. Assuming our object of interest is opaque to thermal radiation, the thermal energy leaving the segmented region consists of energy emitted and reflected from both the fence's surface and surfaces behind the fence but viewed within the gaps between the fence's wood boards. For non-heat generating targets, the amount of energy absorbed, emitted, and reflected from the surface depends on the target's thermal and physical surface properties and amount of energy received by the surface from thermal sources either in front or to the side of the target. The energy received by the target's surface from other thermal sources is called irradiance. The energy leaving the target's surface regardless of the physical cause (emitted plus reflected) is called radiosity. Radiosity is a radiant flux defined as the rate at which thermal radiation leaves the surface due to emission and reflection per unit area of the target's surface ($W \cdot m^{-2}$). Radiosity is the thermal energy received by the detectors in the thermal imaging camera. However, radiosity is normally expressed as radiance ($W \cdot m^{-2} \cdot sr^{-1}$) to associate the quantity to the thermal camera's response displayed by the thermal image, analogous to the human's visual response to luminance [44]. Unless we specify the type of radiance (i.e., emitted or reflected), we will assume all radiance received by the thermal camera is derived from radiosity.

We will now define the terms foreground and background of our thermal scenes with respect to the thermal camera's position and field of view. Foreground is the region in the scene consisting of objects behind the target of interest and within the thermal camera's field of view. Due to the opaqueness of our classes of objects, they are not normally influenced by the thermal radiance emitted from the objects in the foreground. On the other hand, the radiance emitted by the objects in the foreground could have an effect on the thermal camera's AC coupling. As discussed in Chap. 2, AC coupling could result in a target with a constant thermal radiance being assigned variations in gray-level values depending on the radiance of the foreground. Fortunately, these variations in the gray-level values of a target's thermal image will not impact our ability to generate features as long as the AGC is disabled. Background is defined as the region either in front or to the side of the target consisting of thermal sources that emit irradiance onto the target's surface. The source emitting this irradiance may or may not be in the camera's field of view. Referring back to Fig. 3.1, a portion of the total thermal radiance received by the camera comes from the foreground radiance emitted from the gaps in wood fence (denoted by the dotted arrow) and background irradiance from the vinyl siding on the building (dashed arrow) that is both reflected from the fence's surface (dashed arrow) and absorbed and then emitted from the fence's surface (solid arrow).

Figure 3.1 (c) displays the gray-level array (or matrix) of the thermal image segment denoted with the rectangular solid border in Fig. 3.1 (b). The gray-level array consists of M rows and N columns such that each pixel element at coordinate (r, c) is mapped to a gray-level value from the range [0, 255] by the function $I(r, c)$ that depends on the radiance emitted by the surfaces in the thermal image. Thus, a surface emitting a high amount of radiance is assigned a higher gray-level value compared to a surface that is emitting a lower radiance. In the field of thermography for nondestructive testing (NDT) (or nondestructive evaluation (NDE)), a thermal imaging camera is used to record the distribution of apparent surface temperatures to assess the structure or behavior of what is under the surface [32]. To compute these apparent temperatures of the structure's surface, the operator must input the object's emissivity and ambient temperature

[37, 38]. However, in our application we are seeking to assign a class to an unknown object. Therefore, we do not know the target's emissivity. Consequently, we will relate the gray-level values to the amount of radiance emitted by the objects' surfaces in a thermal image, not their apparent surface temperatures. As shown in Fig. 3.1 (c) the values of the gray-levels decrease from left to right indicating a region of higher radiance emission on the left and lower emission on the right side of the segment in Fig. 3.1 (b). Moreover, the object's radiance input is not linearly related to the thermal camera's digital gray-level value output due to AC coupling.

3.4 Meteorological Features

Since the thermal properties (such as conductivity, emissivity, and specific heat) of our non-heat generating objects primarily depend on solar energy, the amount of thermal radiance emitted at the surface is dependent on solar energy as well. Therefore, we can estimate current and historical effects of the diurnal cycle of solar energy on the amount of radiance emitted from an object's surface by generating features based on the ambient temperature.

3.4.1 Ambient Temperature

The effects of solar energy on the amount of radiance emitted from an object's surface is estimated by the ambient temperature ($°F$) feature recorded in same vicinity of the target at the time (t) defined by:

$$Ta = T_a[t] \tag{3.1}$$

3.4.2 Ambient Temperature Rate of Change

The historical effects of solar energy on the amount of radiance emitted from an object's surface is determined by a first order backward difference quotient about the current time (t) with $\Delta t = 30$ minutes.

$$T1 = \frac{T_a[t] - T_a[t - \Delta t]}{\Delta t} \tag{3.2}$$

3.5 Micro Features

Micro features are based on the thermal-physical properties of our targets' surfaces. Particularly, we derive micro features based on the emissivity of an object. The term emissivity is assigned to ideal materials and emittance is used to characterize real materials with surface defects and irregularities [38]. However, we use emissivity for our real materials to avoid confusion since this term is used most often in the infrared community.

Emissivity is a surface property that provides a measure of an object's ability to emit thermal energy. Emissivity is expressed as the ratio of thermal radiation emitted by an object's surface to the thermal radiation emitted by a perfect emitter (blackbody) under the same surface temperatures, viewing angle, and spectral wavelengths [32]. Emissivity is a unitless quantity on a scale from 0 to 1. A perfect emitter of thermal radiation has an emissivity value of unity while a perfect reflector has an emissivity value of zero. When an object is in thermal equilibrium with its local environment, Kirchhoff's law implies that the amount of thermal energy emitted by an object's surface is approximately equal to the amount absorbed for a specified wavelength and direction. Therefore, a common saying in the thermography community is that a good absorber is a good emitter and a poor absorber is a good reflector.

A material's emissivity is not a constant parameter. Emissivity is a function of the type of material, viewing angle, and the object's surface quality, shape, and temperature [32, 38, 39, 40, 45, 46]. Emissivity could also vary with wavelength; however, in our research we assume all objects are graybody emitters. If an object is a graybody emitter, its emissivity will not depend on wavelength [37]. The amount of thermal radiance emitted by a target and detected by a thermal imaging camera depends on the emissivity of the target. Thus, the higher an object's emissivity, the more thermal radiance it will emit.

3.5.1 Emissivity Variation by Material Type

Emissivity varies by the type of material (metallic or nonmetallic) and type of coating on the surface (such as paint, dust, dirt, or corrosion due to oxidation). Polished metallic surfaces generally have a low emissivity (appear very reflective), but the amount of thermal emission can be increased by the presence of certain paints or oxide layers on the surface. As an example of the reflective qualities of a polished metallic surface, consider the visible image of the aluminum plate in Fig. 3.2 (a). The thermal image of the aluminum plate in Fig. 3.2 (b) displays the irradiance from a portion of a house in the background being

56   3 Thermal Feature Generation

reflected off the plate. We may also have to contend with objects that have a high emissivity and are opaque to thermal radiation. For example, consider the plate of glass with an emissivity of approximately 0.92 [38] in front of the pine tree log in Fig. 3.3 (a). As we see in Fig. 3.3 (b), the glass plate is opaque to the thermal radiation emitted by the pine tree log displayed in Fig. 3.3 (c). The emissivity values of metallic and nonmetallic materials are available in many references with topics involving thermography and/or radiative heat transfer [32, 38, 45, 46].

Fig. 3.2 Aluminum plate with low emissivity. (a) visible image of aluminum plate. (b) thermal image of aluminum plate.

3.5 Micro Features 57

Fig. 3.3 Glass plate with high emissivity and opaque to IR radiation. (a) visible image of glass plate in front of pine tree log. (b) thermal image of glass plate in front of log. (c) thermal image of log without glass plate in front.

3.5.2 Emissivity Variation by Viewing Angle

The variation of emissivity with the viewing angle of the thermal camera with respect to the target also depends on the target's surface material. Some typical trends in the emissivity of nonmetallic and metallic materials are shown in Fig. 3.4, as given by [46]. For nonmetallic materials such as wood and vegetation, the emissivity remains rather constant across variations in the viewing angle up to about 50 degrees from normal incidence [32]. On the other hand, the emissivity of smooth metallic surfaces tends to be lower at normal incidence than at other viewing angles.

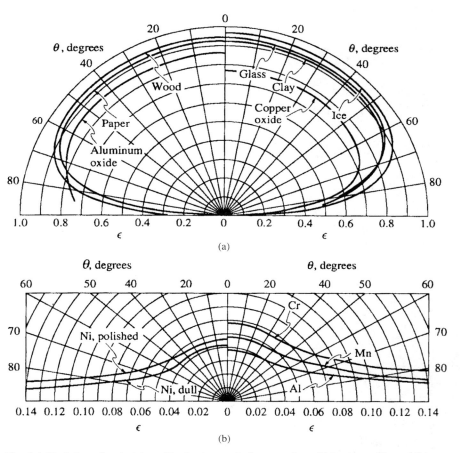

Fig. 3.4 Variation of emissivity with viewing angle for a number of (a) nonmetallic and (b) metallic materials. [46]

3.5.3 Emissivity Variation by Surface Quality

The effects of surface quality on the thermal radiance emitted from a target's surface are difficult to model since the characteristics of smoothness or roughness may be very different from surface to surface. A discussion on models used to measure surface roughness is found in [46]. In general, smooth, polished surfaces like the aluminum plate in Fig. 3.2 (a) can result in a more specular reflection (lower emissivity) than rough surfaces such as bricks that have a diffuse surface (higher emissivity).

3.5.4 Emissivity Variation by Shape and Surface Temperature

When viewing a still frame of a vertically placed cylindrical object with uniform irradiance using a thermal imaging camera, we should witness a variation in radiance as we scan horizontally from the center to the periphery of the object in the image. This variation in radiance is due to the object's directional variation of emissivity. On the other hand, we should not see any significant variation in radiance when scanning the thermal image of a flat object in the same manner with the camera at normal incidence. To demonstrate how emissivity varies with an object's shape (directional variation) and surface temperature, black electrical tape was wrapped around a cardboard cylindrical tube placed in a position such that the irradiance was constant and uniformly distributed. The interior of the tube was heated to $114.8°F$ and thermal images were captured at increments of $2°F$ as the tube cooled to an ambient temperature of $56.3°F$. An averaged vertical radiance (gray-level) was computed using the thermal radiance from the tape in each thermal image. Figure 3.5 displays how the averaged radiance varies horizontally

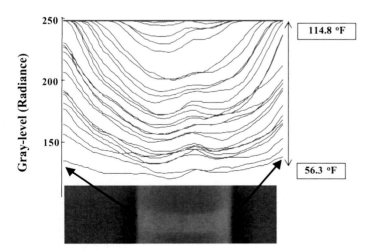

Fig. 3.5 Variation of emissivity with object shape and surface temperature.

60   3 Thermal Feature Generation

along the segment of tape. Since the irradiance is constant and uniformly distributed, the variation in radiance at each temperature increment is due to the directional variation of emissivity. However, we must be aware that a higher surface temperature does not necessarily yield a higher emissivity. The emissivity of a conductor will increase with increasing surface temperature, but the emissivity of a nonconductor may either increase or decrease with increasing surface temperature depending on the specific material [45].

3.5.5 Other Directional Variation Enhancers

Vertical cylindrical objects in an unstructured outdoor environment will not only display variations in radiance due to the directional variation of emissivity, irradiance from the background and solar energy could also have a significant effect on the variation in radiance. Thus, the irradiance as well as the surface temperature may not be uniformly distributed on the object. For instance, consider an experiment to capture a thermal image of a pine tree log with varying irradiance from sources in the background. Thermal images of a pine tree log were captured outside on 29 December 2006 with an ambient temperature of $66.9°F$. The thermal imaging camera captured the images while facing the center segment of the log at normal incidence, 2.4 meters from the log. The surface temperature measurements of the pine tree log at the time the images were captured along with the experimental setup are shown in Fig. 3.6 (a). A building's brick wall with a surface temperature of $80.2°F$ is located 3.4 meters to the left of the log and the sun is located in the direction as displayed in Fig. 3.6 (a). Figure 3.6 (b) shows the thermal image of the log with the irradiance from the brick wall. Figure 3.6 (c) shows the thermal image of the log with the irradiance from the wall blocked using a sheet of drywall positioned 0.6 meters from the log. Figure 3.6 (d) compares the gray-level values as we scan horizontally along the tape segment on the log of the irradiance from the brick wall and the irradiance from the dry wall (brick wall blocked). The scenario presented in Fig. 3.6 allows us to see the simultaneous effects of solar energy, irradiance from the background, and directional variation of the object's surface emissivity. Perhaps we would expect a decrease in the radiance on the left side of the log when the irradiance from the brick wall was blocked using the sheet of drywall. On the contrary, the sheet of drywall introduced a new and greater source of irradiation. However, as we scan from the center to the right of both images, the radiance remains approximately equal.

In Chap. 2, we introduced the halo effect commonly viewed in thermal images where a strong thermal contrast exists between the target's surface and foreground within the camera's field of view. As we discussed, this halo effect is the result of the mechanical chopper wheel within the camera during cyclic process of capturing a thermal image of an object. Two scenarios will result in a halo appearing around an object in its thermal image. First, a "hot" target and "cold" foreground will result in a thermal image with a halo around the "hot" target that has a smaller gray-level value (darker shade) than the "cold" foreground as displayed in Fig. 3.7 (a). The second scenario is a "cold" target and "hot" foreground resulting in a thermal image with a halo around the "cold" target that has a larger gray-level value (lighter shade) than the "hot" foreground as displayed in Fig. 3.7 (b). Consequently, the halo around the target in these two scenarios will also influence how the camera's AC

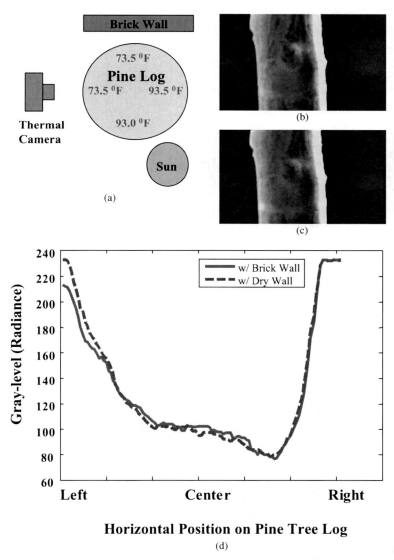

Fig. 3.6 Directional variation of emissivity for a pine tree log outdoors. (a) experimental setup, (b) pine tree log with brick wall irradiance, (c) pine tree log with dry wall irradiance. (d) gray-level comparisons of brick wall vs. dry wall.

coupling will effect the assignment of gray-level values at the periphery in the target's thermal image. Thus, the peripheries of the pine tree log in Fig. 3.7 (a) are assigned a large gray-level value (lighter shade) due to the neighboring halo with a "colder" (smaller gray-level value) apparent temperature than the actual foreground. On the other hand, the peripheries of the pine tree log in Fig. 3.7 (b) are assigned a small gray-level value (darker shade) due to the neighboring halo with a "hotter" (larger gray-level value) apparent temperature than the actual foreground. These two scenarios of the halo effect will also contribute to the variations in radiance that already exist due to the directional variation of emissivity and irradiance from sources in the background.

The majority of our compact objects display variations in radiance from the center to the peripheries due to the directional variation of emissivity, irradiance from sources in the background, and/or halo effect. Since these larger variations

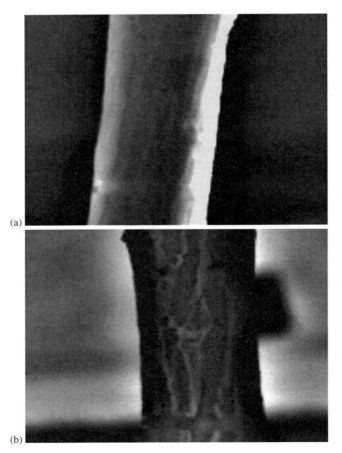

Fig. 3.7 Halo effect resulting from a (a) "hot" target and "cold" foreground and (b) "cold" target and "hot" foreground.

in radiance at the peripheries caused by irradiance from sources in the background and the halo effect may interfere with our ability to generate relevant features, we will generate all our features for compact objects using only their center segment in the thermal image.

3.5.6 Emissivity-based Features

The amount of thermal radiation emitted by our non-heat generating objects will depend on their emissivity and thermal irradiance emitted from external sources in the environment. The primary external source of thermal energy for our outdoor, non-heat generating objects is the sun. Therefore, features based on emissivity will allow us to capture variations in thermal-physical properties that depend on the solar energy and are unique to an object class.

The fundamental equation that allows us to measure the radiance emitted from an object's surface is given by [44]:

$$L_o(T_o) = \tau \varepsilon_o \tilde{L}(T_o) + \tau(1-\varepsilon_o)\tilde{L}(T_b) + (1-\tau)\tilde{L}(T_a) \quad (3.3)$$

where L_o is the radiance detected by the camera, $\tilde{L}$ is the total radiance of a blackbody, T_o is the surface temperature of the object, T_b is the background temperature, T_a is the ambient temperature, τ is the transmission coefficient of the atmosphere, and ε_o is the emissivity of the object (the object is assumed to be a graybody emitter and opaque). Since we are maintaining a camera to target distance of 2.4 m, we can neglect any effects due to the atmosphere and assign an atmospheric transmittance of approximately 100% [32] so that Eq. 3.3 becomes:

$$L_o(T_o) = \varepsilon_o \tilde{L}(T_o) + (1-\varepsilon_o)\tilde{L}(T_b) \quad (3.4)$$

If we assume an opaque object with a diffuse surface, the distribution function $(1-\varepsilon_o)\tilde{L}(T_b)$ is independent of the incidence reflection angles so that $(1-\varepsilon_o)\tilde{L}(T_b) \approx \frac{(1-\varepsilon_o)}{\pi} E$, where E is the irradiance energy on the target from the surrounding background environment. As noted in [44], $\frac{E}{\pi}$ can be evaluated by measuring the radiance reflected by a diffuse surface, such as crinkled aluminum foil. Aluminum foil is a good reflector of thermal radiation due to its low emissivity value (approximately 0.04 for a wavelength of 10 μ m at 78.8 $°F$) [38].

64   3 Thermal Feature Generation

Letting our irradiance from the background be estimated by $L_b(T_b) = \dfrac{E}{\pi}$, Eq. 3.4 becomes:

$$L_o(T_o) = \varepsilon_o \tilde{L}(T_o) + (1-\varepsilon_o) L_b(T_b) \qquad (3.5)$$

The scenario for the radiance received by the thermal imaging camera from an object's surface is displayed in Fig. 3.8 (a). Figure 3.8 (b) shows a thermal image of the cedar tree displayed in Fig. 3.8 (a) captured at 0545 hrs (before sunrise) on

Fig. 3.8 (a) Thermal radiance received by the thermal imaging camera. (b) Thermal image of cedar tree captured at 0545 hrs on 17 March 2006.

17 March 2006. The ambient temperature was approximately 45.7°F. The mobile robot was positioned as displayed in Fig. 3.8 (a). Not only is the thermal imaging camera able to detect radiance coming from the cedar tree but we also see the influence of irradiance coming from the brick wall as indicated by the higher radiance on the right side of the cedar tree in the thermal image.

We will now derive an equation to estimate the emissivity of an object, ε_o, using a reference emitter with a known emissivity of ε_r that is applied to the object so both are at the same surface temperature, subject to the same thermal irradiance from the background, and opaque. Let the radiance from the object's surface be given by Eq. 3.5 and the radiance from the reference emitter's surface be given by:

$$L_r(T_o) = \varepsilon_r \tilde{L}(T_o) + (1-\varepsilon_r) L_b(T_b) \qquad (3.6)$$

We now solve Eqs. 3.5 and 3.6 in terms $\tilde{L}(T_o)$, and algebraically combine the resulting equations to eliminate $\tilde{L}(T_o)$. From Eq. 3.5 we have:

$$\tilde{L}(T_o) = \frac{L_o(T_o) - (1-\varepsilon_o) L_b(T_b)}{\varepsilon_o} \qquad (3.7)$$

From Eq. 3.6 we have:

$$\tilde{L}(T_o) = \frac{L_r(T_o) - (1-\varepsilon_r) L_b(T_b)}{\varepsilon_r} \qquad (3.8)$$

Combining these results we have:

$$\frac{L_o(T_o) - (1-\varepsilon_o) L_b(T_b)}{\varepsilon_o} = \frac{L_r(T_o) - (1-\varepsilon_r) L_b(T_b)}{\varepsilon_r} \qquad (3.9)$$

Solving for ε_o we obtain our desired equation for the emissivity:

$$\varepsilon_o = \frac{L_o(T_o) - L_b(T_b)}{L_r(T_o) - L_b(T_b)} \varepsilon_r \qquad (3.10)$$

Madding [37] uses this result to investigate how emissivity measurement accuracy affects temperature measurement accuracy.

For our micro features, we continue to recognize the surface radiances' dependencies on temperature; however, we simplify our emissivity equation by letting $L_o = L_o(T_o)$, $L_r = L_r(T_o)$, and $L_b = L_b(T_b)$. To ensure consistency with the notation used for our thermal-physical features, we will also change our symbol for emissivity so that $Eo = \varepsilon_o$. Therefore, our emissivity feature is defined by:

$$Eo = \frac{L_o - L_b}{L_r - L_b} \varepsilon_r \qquad (3.11)$$

66   3 Thermal Feature Generation

To compute the emissivity feature Eo, the values of L_0, L_r, and L_b are first derived from the mean of the thermal radiance (gray-level values) of surface segments in the thermal images of the object, reference emitter, and aluminum foil, respectively. These values are then substituted into Eq. 3.11 to obtain our estimate of Eo. As noted earlier, emissivity is a function of the type of material, viewing angle, and the object's surface quality, shape, and temperature. Since compact objects (particularly cylindrical objects) display variations in radiance from the center to the peripheries due to the directional variation of emissivity, irradiance from sources in the background, and halo effect, the emissivity feature was computed using the center image segment on all compact objects. For thermal scenes of extended objects that lack thermal emissions from a foreground, such as dense hedges and brick walls, the surface segment used to compute L_0 consists of all the constituents that make up the object. For instance, the segment selected on the

Fig. 3.9 Visible and thermal images of objects captured on 10 February 2007 to evaluate the emissivity feature. (a) steel pole, (b) birch tree log, (c) concrete cylinder, (d) hedges, and (e) wood wall.

hedges to compute L_0 primarily consists of leaves but also includes branches. The segment selected on brick walls to compute L_0 consists of the brick and the mortar between the bricks. For thermal scenes of extended objects that display a thermal radiance from the foreground, such as wood walls and picket fences, only a segment of the extended object's surface is selected in the image to compute L_0. Crinkled aluminum foil with an emissivity of approximately 0.04 [38] was attached to the target afterwards to compute the irradiance energy on the object from the surrounding background environment, L_b. The aluminum foil must not be attached to the target prior to capturing the thermal image to compute L_0 in order to avoid disturbing the natural radiance being emitted by the target. The reference emitter was black electrical tape attached to the object with a known emissivity ε_r of approximately 0.97 [38]. The black electrical tape should be attached to the surface of the target well in advance in capturing thermal images of the target to ensure the tape obtains the same surface temperature as the target. The segmented region of the target used to compute L_0 does not include the reference emitter.

As an example of our emissivity feature, thermal images of a steel pole, birch tree log, concrete cylinder, hedges, and wood wall (see Fig. 3.9) were captured at various times on 10 February 2007. The black electrical tape used as the reference emitter is shown attached to the targets in each thermal image. All thermal images were captured as described in Chap. 2 with a distance of 2.4 meters between the Raytheon *ControlIR* 2000B long-wave infrared thermal imaging video camera and the object. The thermal images were captured with the thermal camera facing the center of each object at normal incidence. Table 3.1 provides the ambient temperatures of the environment and surface temperatures of the objects at the times the thermal images were captured. The average ambient temperatures are noted in Table 3.1 for each time interval. The surface temperatures of the objects were recorded at the time the thermal image was captured. All objects were influenced by the same solar conditions during each time interval.

Table 3.2 provides the generated feature values for the objects at the times the thermal images were captured. By analyzing Table 3.2, we notice trends in the emissivity feature values that allow us to distinguish one object from another. Furthermore, a detailed analysis of both Tables 3.1 and 3.2 reveals how the emissivity feature lets us also consider the effects of other thermal properties. For instance, emissivity depends on surface temperature (as well as the type of material, viewing angle, and the object's surface quality and shape) and surface temperature depends

Table 3.1 Thermal image capture times and temperatures for objects in Fig. 3.9 captured on 10 February 2007.

Time	Ambient Temp (°F)	Surface Temp (°F)				
		Steel Pole	Birch Log	Concrete Cyl	Hedges	Wood Wall
0745	25.7					
0900	37.5					
1000(+/− 15min)	45.8	545	49.4	50.7	45.1	47.4
1330(+/− 15 min)	46.5	543	53.2	54.8	48.2	62.0
1615(+/− 15 min)	42.2	512	53.6	48.7	43.1	52.5

Table 3.2 Feature values generated from the thermal image of objects in Fig. 3.9 captured on 10 February 2007.

Time (+/−15min)	Ambient Temp (°F) (Ta)	Ambient Temp Rate of Change (T1)	Object	Emissivity (Eo)
1000	45.8	0.04	Steel Pole	0.8876
			Birch Log	0.3106
			Concrete Cyl	0.1922
			Hedges	−0.0912
			Wood Wall	0.4623
1330	46.5	0.02	Steel Pole	0.8792
			Birch Log	0.4498
			Concrete Cyl	0.4187
			Hedges	−0.0477
			Wood Wall	0.0320
1615	42.2	−0.03	Steel Pole	0.7803
			Birch Log	0.3581
			Concrete Cyl	0.4564
			Hedges	−0.2772
			Wood Wall	−0.0958

on the specific heat (as well as conductivity and other thermal properties) of the object. The surface temperature of low-specific-heat objects, such as the leaves on the hedges, tend to track the availability of solar energy [38]. When a cloud passes or the sun begins to set, the surface temperature of the hedges stays consistent with the lower ambient temperature. Moreover, a low level of solar energy available to a low specific heat object results in less thermal radiation emitted as indicated by the hedges' consistently low emissivity presented in Table 3.2. On the other hand, objects with a high specific heat, such as the birch tree log ($\sim 2.4\ kJ \cdot kg^{-1} \cdot {}^{\circ}C^{-1}$) [32], will tend to heat up more slowly with the increasing solar energy and cool more slowly as the amount of solar energy begins to decrease in the late afternoon (around 1600 hrs). The emissivity of the birch tree log first increases with the availability of solar radiation in the morning as indicated by the positive rate of change in ambient temperature in the morning. As the solar energy decreases throughout the afternoon, the emissivity of the birch tree log slightly lowers in value as expected. Along with the possibility of some error in the temperature measurement, we see no significant change in the surface temperature of the birch tree log between 1330 and 1615 hrs due to the effect of its specific heat. Even though the steel pole has a low specific heat ($\sim 0.47\ 2.4\ kJ \cdot kg^{-1} \cdot {}^{\circ}C^{-1}$) [32], its emissivity consistently shows the highest value due to the light coating of black paint ($\varepsilon \sim 0.96$ at $75.2°F$ in a controlled environment) [38] and oxidation on the surface. An interesting observation is that the black electrical tape used as the reference emitter attached to the steel pole (Fig. 3.9 (a)) emits a slightly higher radiance than the steel pole since the tape's emissivity is approximately 0.97.

We also notice that our emissivity values do not necessarily vary between 0 and 1 as is the case of experiments in a controlled inside laboratory environment. By observing Eq. 3.11, we see that the emissivity values could be quite sensitive to variations in the thermal radiance of the object, reference emitter, and aluminum foil. For instance, as the radiance of the reference emitter and the aluminum foil approach the same value, the denominator in the equation for emissivity will become very small (either positive or negative). As a result, the value of the emissivity in Eq. 3.11 would take on very large values (either positive or negative). We will illustrate in Chap. 4 that these extreme value of emissivity are rare and will be treated as outliers. To avoid such extreme feature values, we use the following additional micro features derived from the emissivity given in Eq. 3.11:

$$Lo = L_o \tag{3.12}$$

$$Lr = L_r \tag{3.13}$$

$$Lb = L_b \tag{3.14}$$

$$Lor = \frac{L_o}{L_r} \tag{3.15}$$

$$Lob = \frac{L_o}{L_b} \tag{3.16}$$

Lr and Lb are only used in conjunction with features generated from the thermal radiance emitted from the target and not used to discriminate targets as stand-alone features. The features Lor and Lob were chosen to create a ratio value. Other types of features could be used as well; however, additional choices, such as $L_o - L_r$ or $L_o - L_b$, will more likely have a strong correlation with our existing features and result in redundancy in the feature set.

3.6 Macro Features

Macro features provide a unique representation of a target based on the spatial variation in radiance (gray-level values) observed in the thermal image. Macro features seek to generate descriptors that not only consider radiant patterns found on the target's surface but also patterns observed in the entire thermal image of the target within the camera's field of view. Thus, macro features may also consider patterns formed by gaps in the target that allow the camera to receive radiation emitted from the foreground. For instance, macro features allow us to generate features that describe the periodic pattern of wood boards on the fence in Fig. 3.1 (b). Since compact objects (particularly cylindrical objects) display variations in radiance from the center to the peripheries due to the directional variation of emissivity, irradiance from sources in the background, and halo effect, we will always compute the macro features using the center image segment on all compact objects. On the other hand, for the extended objects, we will compute their macro features using the entire scene within the camera's field of view. Our macro features are derived from first- and second-order texture features.

3.6.1 First-order Statistical Features

First-order statistics provide measures based on the probability of observing a gray-level value at a random location in the thermal image. Our first-order statistics are generated using a histogram of pixel intensities from an object's thermal image. Our histograms and first-order statistics follow from those presented in [2]. The histogram of each thermal image has a total of 256 possible intensity levels in the interval [0, 255] defined as a discrete function:

$$h(r_k) = n_k \tag{3.17}$$

where r_k is the kth intensity level on the interval [0, 255] and n_k is the number of pixels in the thermal image that have an intensity level of r_k. The kth indices take on values from 1 to 256 associated with the position of the gray-level value in [0, 255]. The probability $P(r_k)$ of observing a gray-level value at a random location in the thermal image is given by the normalized form of the histogram:

$$P(r_k) = \frac{h(r_k)}{n} = \frac{n_k}{n} \tag{3.18}$$

where n is the total number of pixels in the thermal image. With this convention, we now define our first-order statistics.

3.6.1.1 Object Scene Radiance

The object scene radiance is the average of the radiance coming from the target's surface and any foreground emitters within the field of view of the segmented target. The mean for the first-order statistics is defined as:

$$Mo1 = \sum_{k=1}^{256} r_k \, P(r_k) \tag{3.19}$$

3.6 Macro Features

The following two variations of $Mo1$ were used to consider the radiance emitted by the reference emitter and background, respectively:

$$Mor1 = \frac{Mo1}{L_r} \quad (3.20)$$

$$Mob1 = \frac{Mo1}{L_b} \quad (3.21)$$

Since $Lo = Mo1$ for compact objects, $Mo1$, $Mor1$, and $Mob1$ only apply to extended objects.

The mean radiance can also be used to generate texture features based on the nth moment about the mean $Mo1$:

$$\mu_n = \sum_{k=1}^{256} (r_k - Mo1)^n P(r_k) \quad (3.22)$$

However, we limit our moments to order $n = 3$ so that our features maintain their physical interpretations. The following two features are based on the second and third moments, respectively.

3.6.1.2 Contrast1

Contrast is a measure of the amount of variation in the radiance of an object in a thermal image. The contrast feature is based on the standard deviation of the gray-level values about the mean $Mo1$ given by:

$$Co1 = \sqrt{\sum_{k=1}^{256} (r_k - Mo1)^2 P(r_k)} \quad (3.23)$$

3.6.1.3 Smoothness

Smoothness measures the variations in the intensity of the gray-level values of an object's thermal image as computed by:

$$So1 = 1 - \frac{1}{(1 + Co1^2)} \quad (3.24)$$

Values of $So1$ close to zero represent surfaces with a constant gray-level value and values close to unity imply surfaces with large deviations among their gray-level values.

3.6.1.4 Third Moment

The third moment is defined by:

$$Tol = \sum_{k=1}^{256} (r_k - Mo1)^3 P(r_k) \qquad (3.25)$$

The third moment measure the skewness of the distribution of gray-level values in the histogram. When the histogram is symmetric, the value of the third moment is zero. When the histogram is skewed to the right or left about the mean, the value of the third moment is accordingly positive or negative, respectively.

3.6.1.5 Uniformity

The uniformity feature is defined by:

$$Uo1 = \sum_{k=1}^{256} \left[P(r_k) \right]^2 \qquad (3.26)$$

The value of uniformity increases as the histogram of gray-level values approaches a uniform distribution and is unity for a thermal image of an object with a constant surface radiance.

3.6.1.6 Entropy1

The entropy feature provides a measure of randomness (or complexity) in the intensity (gray-level) values of an object's thermal image. The use of the term entropy can easily cause some confusion since there are continuous debates within the scientific community concerning the correct definition of entropy. Therefore, before we present our use of entropy and derive an equation for the term, we will first provide some background information on entropy.

The term entropy was first introduced in classical thermodynamics. However, the definition has become rather subjective to fit the needs of other fields of study. Thus, one can find different definitions in thermodynamics, chemistry, information theory, and other fields. For instance, a search on the internet results in the following definitions: entropy is a measure of randomness; entropy is a measure of the probability of a particular result; entropy is a measure of the disorder of a system; entropy measures the heat divided by the absolute temperature of a body. Some of the names associated with the definition of entropy include Clausius, Gibbs, Boltzmann, Szilard, von Neumann, Shannon, and Jaynes. Shannon was interested in communication theory and von Neumann investigated quantum mechanical entropy. Shannon initiated the use of the quantity $H = -K\sum P_i \log P_i$ (where K is a positive constant) in information theory as a measure of "information, choice, and uncertainty" [47]. However, regarding a name for H, Shannon stated [48]:

> My greatest concern was what to call it. I thought of calling it 'information,' but the word was overly used, so I decided to call it 'uncertainty.' When I discussed it with John von Neumann, he had a better idea. Von Neumann told me, 'You should call it entropy, for two reasons. In the first place your uncertainty function has been used in statistical mechanics under that name, so it already has a name. In the second place, and more important, no one knows what entropy really is, so in a debate you will always have the advantage.'

As a result, Shannon's entropy was introduced in information theory. In [47], Shannon states, "In the discrete case the entropy measures in an absolute way the randomness of the chance variable."

The next step is to find a definition of entropy that is applicable to classifying objects in thermal imaging application. An appropriate definition for entropy is found in the digital image processing community in the area of texture analysis and pattern classification [2, 3, 4, 42, 49, 50, 51]. The entropy used in digital image processing is consistent with Shannon. In digital image processing, entropy is defined as a statistical measure of randomness in the intensity values of an object's visible image, and used to characterize the texture of objects in an image [51]. For our application, we adopt the same definition; however, we measure the randomness in the intensity (gray-level) values pertaining to an object's thermal image. From this definition, we derive our equation for the entropy feature.

Following the mathematical framework of information theory, our measure of randomness in the gray level values is given by:

$$R(r_k) = -\log_2(P(r_k)) \qquad (3.27)$$

where the choice of the base is consistent with units, in bits, for measuring information. Consequently, if only one gray level value, say r_1, was present in the thermal image, $P(k_1) = 1$ and $R(k_1) = 0$ so no randomness would occur. From Eq. 3.17, we have n_k cases with randomness measure $R(r_k)$, the average value of randomness in our object's thermal image follows from Eqs. 3.18 and 3.27 as:

$$\frac{\sum_{k=1}^{256} n_k R(r_k)}{n} = \sum_{k=1}^{256} \frac{n_k}{n}(-\log_2(P(r_k)))$$

$$= -\sum_{k=1}^{256} P(r_k) \log_2(P(r_k)) \qquad (3.28)$$

This last quantity, called the entropy, will provide our required measure of randomness in the gray-level values of an object's thermal image. Therefore, our entropy feature value is computed by:

$$En1 = -\sum_{k=1}^{256} P(r_k) \log_2 (P(r_k)) \qquad (3.29)$$

where $En1$ increases in value as the randomness in the gray-level values increases in the object's thermal image.

3.6.2 Second-order Statistical Features

Second-order statistics methods also provide a way to generate features that describe the radiant patterns in the thermal image of an object. Thus, second-order statistics features are our second type of macro features. However, unlike first-order statistical methods that depend only on individual gray-level values, second-order statistical methods involve the interaction or co-occurrence of neighboring gray-level values. Second-order statistics provide measures based on the probability of observing pairs of gray-level values with a defined spatial relationship in an object's thermal image. The spatial relationship consists of a specified direction and distance between a pair of gray-level values. The macro features are generated from the spatial relationships that are reported in a gray-level co-occurrence matrix (GLCM), also known as a gray-level spatial dependence matrix. Our second-order statistical features follow from those presented in [51] and are based on the pioneering work of Haralick, Shanmugam, and Dinstein [49]. Other notable discussions on second-order statistical features involving the GLCM are found in [42, 50].

The GLCM records how often a pixel of interest with a gray-level value of i occurs in a specific spatial relationship to a pixel with a gray-level value of j in a thermal image. A pixel of interest in a thermal image forms a spatial relationship with one of its neighboring pixels defined by a pixel distance D and direction (angle) denoted by a row vector with the pixel of interest as the origin as illustrated in Fig. 3.10 (a). We choose four directions (0°, 45°, 90°, and 135°) to afford our macro features the ability to capture discriminating information along various directions on a target's surface. Our choice of angles assumes that the thermal radiant patterns are symmetric along each direction about the pixel of interest. The most favorable pixel distance D is the one that allows a spatial relationship that captures an object class's distinctive radiant patterns. We will discuss our most favorable pixel distances for both extended and compact objects after we present our second-order features below.

Suppose Fig. 3.10 (b) illustrates a gray-level array of a thermal image with gray-level values ranging from 0 to 3. The four GLCMs for each direction and a distance $D = 1$ are provided in Figs. 3.10 (c–f). The shaded regions in each GLCM displays

3.6 Macro Features 75

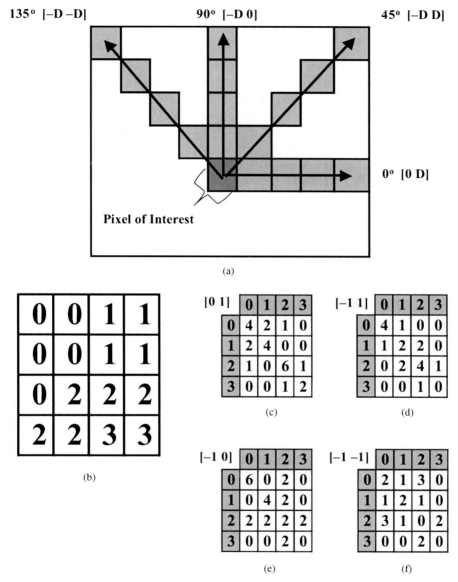

Fig. 3.10 Gray-level Co-occurrence Matrix. (a) spatial relationship of neighboring pixels, (b) gray-level array of a thermal image, (c)–(f) GLCMs with distance $D = 1$ and directions 0, 45, 90, and 135 degrees, respectively.

the gray-level values of the pixel of interest (i) along the first column and its neighboring pixel's gray-level values (j) along the first row. As we see, the number of gray-level values in the thermal image determines the size of the GLCM. Each element

(i, j) in the GLCM provides the number times that a pixel with gray-level value i occurred in the specified spatial relationship with the pixel with gray-level value j in the thermal image. We denote this frequency by $f(i, j)$. For example, (1, 0) in Fig. 3.10 (c) presents $f(1,0) = 2$ as the number of times that the pixel of interest with gray-level value $i = 1$ occurred at an angle of zero degrees and distance of one pixel away from a pixel with gray-level value $j = 0$. Let R denote the sum of all the frequencies $f(i, j)$ in the GLCM for a specified spatial relationship. For a GLCM defined by a particular spatial relationship, the probability of observing a pixel of interest with a gray-level value of i in a specific spatial relationship to a pixel with a gray-level value of j in a target's thermal image is given by:

$$P(i, j) = \frac{f(i, j)}{R} \qquad (3.30)$$

Equation 3.30 is used to define the following second-order macro features. For each thermal image of an object, four GLCMs are created where each matrix is defined by a specified relationship (a distance and one of the four angular directions). For each second-order feature, feature values are generated for all four GLCMs. The resulting four feature values are averaged to ensure invariance under rotation as suggested in [49].

3.6.2.1 Contrast2

The contrast feature (also known as inertia) is a measure of the amount of radiant variations between a pixel and its specified neighbor over the entire thermal image. A thermal image with a large amount of radiant variations will have a higher value for the contrast feature compared to a thermal image with a small amount of radiant variations. In terms of the GLCM, contrast is a measure of the spread of $P(i, j)$ values about the main diagonal of the matrix. Contrast becomes larger in value with larger values of $P(i, j)$ spreading away from the main diagonal. The contrast feature value is zero for a thermal image of an object with a constant thermal radiance (gray-level value) across its surface. Contrast2 is defined as:

$$Co2 = \sum_i \sum_j |i - j|^2 \, P(i, j) \qquad (3.31)$$

3.6.2.2 Correlation

Correlation provides a measure of linear-dependencies between the gray-level value of the pixel of interest and its specified neighbor over the entire image. The directions in a thermal image consisting of a linear structure will have either a correlation value closer to 1 (positively correlated) or −1 (negatively correlated). On the other hand, an uncorrelated image with a lack of linear structure and/or high amount of noise will result in a correlation value closer to zero. The correlation value for an image with a constant thermal radiance across the surface is undefined. The correlation feature is defined by:

$$Cr2 = \sum_i \sum_j \frac{(i-\mu_x)(j-\mu_y)P(i,j)}{\sigma_x \sigma_y} \tag{3.32}$$

where μ_x and σ_x are the mean and standard deviation of the rows sums of the GLCM formed by $P(i,j)$ and μ_y and σ_y are the statistics of the column sums.

3.6.2.3 Energy

Energy (also known as angular second moment) measures the uniformity of the gray-level values in a thermal image. In a uniform image there are very few intense gray-level transitions between the neighboring pixels. The values of energy become larger as the GLCM has fewer entries of large $P(i,j)$. Such a case exists when the probabilities $P(i,j)$ are clustered near the main diagonal of the GLCM. The energy is unity for a thermal image of an object with a constant surface radiance. On the other hand, the values of energy approach zero as all $P(i,j)$ become more equal in value. The energy feature is defined by:

$$Er2 = \sum_i \sum_j [P(i,j)]^2 \tag{3.33}$$

3.6.2.4 Homogeneity

Homogeneity is similar to the energy feature. The values of homogeneity become larger as larger values of $P(i,j)$ become clustered near the main diagonal of the GLCM. Homogeneity approaches zero as the values of $P(i,j)$ become more equal and spread away from the main diagonal, and is unity for a diagonal GLCM. Homogeneity is defined by:

$$Ho2 = \sum_i \sum_j \frac{P(i,j)}{1+|i-j|} \tag{3.34}$$

3.6.2.5 Entropy2

Similar to the case in first-order statistics, entropy in second-order statistics is a measure of the complexity (or randomness) in the thermal image. A thermal image become more complex as all the values of $P(i,j)$ in the GLCM approach equality, resulting in a larger entropy. Entropy2 is defined by:

$$En2 = -\sum_i \sum_j P(i,j) \log_2(P(i,j)) \tag{3.35}$$

3.6.2.6 Most Favorable Pixel Distances

As we mentioned previously, a pixel of interest in a thermal image forms a spatial relationship with one of its neighboring pixels defined by a pixel distance D and angular direction denoted by a row vector with the pixel of interest as the origin as illustrated in Fig. 3.10 (a). In this section, we will discuss our most favorable pixel distances for both extended and compact objects. The most favorable pixel distance D is the one that allows a spatial relationship that captures an object class's distinctive radiant patterns. We will analyze various distances applied to the thermal images of extended and compact objects captured with approximately the same environmental conditions and location on 27 March 2007 between 1230 and 1300 hrs. The thermal images of the objects were captured during a period where there was a low thermal contrast in the scenes. These conditions will allow us to choose D values for both extended and compact objects that are sensitive to radiant patterns in a thermal image where a low thermal contrast exists. We will proceed to choose our D values by considering the extended and compact objects in separate cases. The methodology for each case consists of first generating the second-order statistical features from GLCMs with spatial relationships with a horizontal angular direction and varying pixel distances D from 1 to 100, $\{[0,D] | D = 1,...,100\}$. Next, we will compare the feature values and choose the D value that results in the greatest distinction the object classes.

The extended objects used in our analysis to choose the most favorable pixel distance D consist of the brick wall, hedges, picket fence, and wood wall displayed in Fig. 3.11. As we see in Fig. 3.12, Energy and Entropy2 provide the best separation of the object classes. Based on these results, we derive an equation that will assist us in choosing the pixel distance that maximizes the discrimination between the object classes. This equation is defined as the absolute sum of the differences in object class feature values as a function of pixel distance given by:

$$Feat\ Diff(D) = \left| \begin{array}{l} (Picket\ F(D) - Hedges(D)) + (Picket\ F(D) - Brick\ W(D)) \\ + (Picket\ F(D) - Wood\ W(D)) + (Hedges(D) - Brick\ W(D)) \\ + (Hedges(D) - Wood\ W(D)) + (BrickW(D) - Wood\ W(D)) \end{array} \right| \tag{3.36}$$

Fig. 3.11 Visible and thermal images of extended objects used for pixel distance analysis and selection. (a) brick wall, (b) hedges, (c) picket fence, and (d) wood wall.

80   3 Thermal Feature Generation

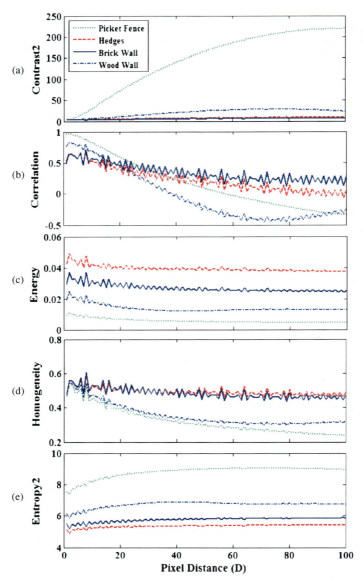

Fig. 3.12 Extended objects pixel distance analysis. Pixel Distance vs. (a) Contrast2, (b) Correlation, (c) Energy, (d) Homogeneity, (e) Entropy2.

By applying this equation to the Energy and Entropy2 features, we obtain the results displayed in Fig 3.13. The pixel distances that provide the best object class separation for Energy is $D = 8$ and Entropy2 is $D = 56$. Comparing these pixel distances to each result in Fig. 3.12, we see that a pixel distance $D = 8$ provides an acceptable separation between the object classes for energy. However, a pixel dis-

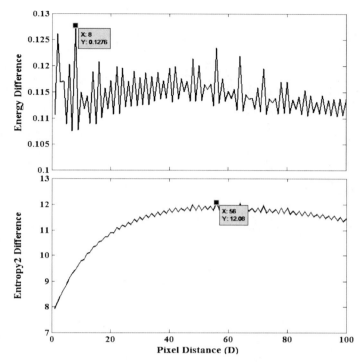

Fig 3.13 Extended objects absolute sum of the differences for Energy and Entropy2 features as a function of pixel distance (D).

tance of $D = 8$ does not result in an acceptable object class separation for the other features. On the other hand, the pixel distance of $D = 56$ for Entropy2 results in an acceptable object class separation for all the second-order statistical features. Consequently, we choose $D = 56$ as the most favorable pixel distance for each spatial relationship involving extended objects.

The compact objects used in our analysis to choose the most favorable pixel distance D consist of the steel poles and trees displayed in Fig. 3.14. As we see in Fig. 3.15, Energy and Entropy2 provide the best separation of the object classes. As with the extended objects we define an equation that will assist us in choosing the pixel distance that maximizes the discrimination between the object classes. However, since we desire to distinguish steel poles from trees for our compact object classes, our equation is given below as the absolute difference of the mean feature values for the three steel poles and three trees across all pixel distances:

$$FeatDiff(D) = \left| \left[\frac{(BrownSteelP(D) + GreenSteelP(D) + OctagonSteelP(D))}{3} \right] - \left[\frac{(BasswoodT(D) + BirchT(D) + CedarT(D))}{3} \right] \right| \quad (3.37)$$

Fig. 3.14 Visible and thermal images of compact objects used for pixel distance analysis and selection. (a) brown steel pole, (b) green steel pole, (c) octagon steel pole, (d) basswood tree (e) birch tree, (f) cedar tree.

By applying this equation to the Energy and Entropy2 features, we obtain the results displayed in Fig. 3.16. Once again, we will choose the pixel distance that maximizes the discrimination between the object classes. The pixel distances that provide the best object class separation for Energy is $D = 8$ and Entropy2 is $D = 16$. Comparing these pixel distances to each result in Fig. 3.15, we see that a pixel distance $D = 8$ provides an acceptable separation between the steel pole and tree object classes for energy. However, a pixel distance of $D = 8$ does not result in an acceptable object class separation for the other features. On the other hand, the pixel distance of $D = 16$ for Entropy2 results in an acceptable object class separ-

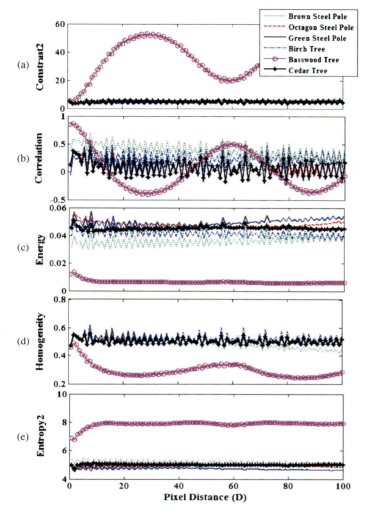

Fig. 3.15 Compact objects pixel distance analysis. Pixel Distance vs. (a) Contrast2, (b) Correlation, (c) Energy, (d) Homogeneity, (e) Entropy2.

ation for all the second-order statistical features. Consequently, we choose $D = 16$ as the most favorable pixel distance for each spatial relationship involving compact objects.

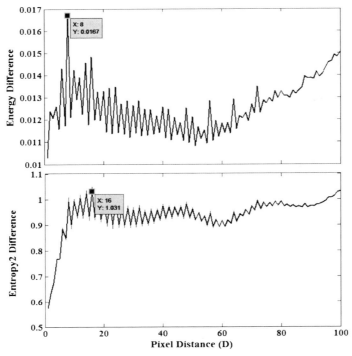

Fig. 3.16 Compact objects absolute sum of the differences for Energy and Entropy2 features as a function of pixel distance (D).

3.7 Thermal Feature Application

We now provide an application to analyze some of the characteristics of our thermal features. However, we will not make any judgments regarding the worthiness of our thermal features. A proper selection of a set of most favorable features will require an exhaustive search using a high performance computing system to analyze the classification performance of every possible combination of features across multiple dimensions. During our exhaustive search, we eliminate redundant features and only retain those sets of features that enhance our ability to distinguish object classes. We delay this exhaustive search until the next chapter. Figure 3.17 displays the thermal images of extended objects (brick wall, hedges, and wood wall) and compact objects (concrete cylinder, steel pole, and pine tree log) that were captured between 0930 and 1400 on 10 February 2007 under approximately the same solar conditions and location. All thermal images were captured as described in Chap. 2 at normal incidence with a distance of 2.4 meters between the Raytheon *ControlIR* 2000B long-wave infrared thermal imaging video

3.7 Thermal Feature Application   85

Fig. 3.17 Visible and thermal images of objects used to evaluate thermal features. Extended objects: (a) brick wall, (b) hedges, (c) wood wall. Compact objects: (d) concrete cylinder, (e) steel pole, (f) pine tree log.

camera and the object. The thermal features were generated on segments of these extended and compact objects using the equations derived in Sects. 3.4–3.6 and summarized in Table 3.3. The resulting feature values are presented in Table 3.4.

Since we intend to distinguish the object classes within either the category of extended or compact objects, we will analyze the two categories separately as disjoint sets of object classes. Beginning with the meteorological features in Table 3.4, we see that the object classes within each category are experiencing approximately the same ambient temperatures and temperature rates of change. In the micro features, the object classes within each category are also experiencing about the same background irradiance. However, the wood wall and pine tree log are both emitting a higher surface radiance compared to the other object classes within their respective category. This higher radiance is partially due to the higher

Table 3.3 Summary of meteorological, micro, and macro features.

FEATURE	EQUATION	FEATURE	EQUATION
Meteorological Features		**Macro Features**	
Ambient Temp. °F (Ta)	$Ta = T_a[t]$	**First-order Statistics**	
Amb. Temp. Rate of Change (T1)	$T1 = \dfrac{T_a[t] - T_a[t-\Delta t]}{\Delta t}$	Object Scene Radiance (Mo1)	$Mo1 = \sum\limits_{k=1}^{256} r_k\, P(r_k)$
Micro Features			
Object Surface Radiance (Lo)	$Lo = L_o$	Mo1/Lr (Mor1)	$Mor1 = \dfrac{Mo1}{L_r}$
Reference Emitter Radiance (Lr)	$Lr = L_r$	Mo1/Lb (Mob1)	$Mob1 = \dfrac{Mo1}{L_b}$
Background Irradiance (Lb)	$Lb = L_b$	Contrast1 (Co1)	$Co1 = \sqrt{\sum\limits_{k=1}^{256}(r_k - Mo1)^2 P(r_k)}$
Lo/Lr (Lor)	$Lor = \dfrac{L_o}{L_r}$	Smoothness (So1)	$So1 = 1 - \dfrac{1}{(1 + Co1^2)}$
Lo/Lob (Lob)	$Lob = \dfrac{L_o}{L_b}$	Third Moment (To1)	$To1 = \sum\limits_{k=1}^{256}(r_k - Mo1)^3 P(r_k)$
Emissivity (Eo)	$Eo = \dfrac{L_o - L_b}{L_r - L_b}\,\varepsilon_r$	Uniformity (Uo1)	$Uo1 = \sum\limits_{k=1}^{256}[P(r_k)]^2$
		Entropy1 (En1)	$En1 = -\sum\limits_{k=1}^{256} P(r_k)\log_2(P(r_k))$
		Second-order Statistics	
		Contrast2 (Co2)	$Co2 = \sum\limits_i \sum\limits_j |i-j|^2 P(i,j)$
		Correlation (Cr2)	$Cr2 = \sum\limits_i \sum\limits_j \dfrac{(i-\mu_x)(j-\mu_y)P(i,j)}{\sigma_x \sigma_y}$
		Energy (Er2)	$Er2 = \sum\limits_i \sum\limits_j [P(i,j)]^2$
		Homogeneity (Ho2)	$Ho2 = \sum\limits_i \sum\limits_j \dfrac{P(i,j)}{1+|i-j|}$
		Entropy2 (En2)	$En2 = -\sum\limits_i \sum\limits_j P(i,j)\log_2(P(i,j))$

specific heat of the wood. Additionally, differences in the radiance are attributed to other factors such as the type of material (including chemicals used on the pressure treated wood wall) and the object's surface quality (smooth vs. rough). Of course these factors also influence the feature values for emissivity. As expected, the wood wall has a higher emissivity compared to the brick wall and hedges. Within the compact objects category, the pine tree log has a median value on the emissivity scale; however, the steel pole has a higher emissivity primarily due to its coating of black paint on the surface.

Table 3.4 Feature values generated from the thermal image of objects in Fig. 3.17.

	Extended Objects			Compact Objects		
	Brick Wall	Hedges	Wood Wall	Concrete Cyliner	Steel Pole	Pine Tree Log
Meteorological Features						
Ambient Temp. $°F$ (Ta)	43.1000	42.2000	46.5000	43.3000	43.3000	46.2000
Amb. Temp. Rate of Change (T1)	0.0367	0.0433	0.0567	0.0100	0.0100	0.0333
Micro Features						
Object Surface Radiance (Lo)	94.8974	94.3022	98.2153	97.0469	100.7415	110.8171
Reference Emitter Radiance (Lr)	119.3813	155.3111	124.6449	109.8459	104.1326	128.6190
Background Irradiance (Lb)	94.4367	97.2481	94.4559	86.7990	87.4212	86.3406
Lo/Lr (Lor)	0.7949	0.6072	0.7880	0.8835	0.9674	0.8616
Lo/Lob (Lob)	1.0049	0.9697	1.0398	1.1181	1.1524	1.2835
Emissivity (Eo)	0.0179	–0.0492	0.1208	0.4313	0.7732	0.5616
Macro Features						
First-order Statistics						
Object Scene Radiance (Mo1)	94.8051	94.2897	94.3128	NA	NA	NA
Mo1/Lr (Mor1)	0.7941	0.6071	0.7567	NA	NA	NA
Mo1/Lb (Mob1)	1.0039	0.9696	0.9985	NA	NA	NA
Contrast1 (Co1)	4.1440	5.1906	12.7304	2.7770	1.6671	6.7767
Smoothness (So1)	0.0003	0.0004	0.0025	0.0001	0.0000	0.0007
Third Moment (To1)	–0.0009	0.0022	–0.0379	–0.0001	0.0000	0.0002
Uniformity (Uo1)	0.1000	0.0716	0.0391	0.1140	0.2065	0.0567
Entropy1 (En1)	3.7119	4.1787	5.2090	3.3448	2.5535	4.6228
Second-order Statistics						
Contrast2 (Co2)	35.4270	53.0582	267.4656	10.6734	3.9182	66.5163
Correlation (Cr2)	–0.0161	–0.0243	0.2170	0.3161	0.2670	0.1995
Energy (Er2)	0.0114	0.0053	0.0031	0.0144	0.0471	0.0039
Homogeneity (Ho2)	0.3358	0.2743	0.2719	0.4134	0.5384	0.2520
Entropy2 (En2)	7.2262	8.2843	9.4353	6.5705	4.9302	8.8415

By analyzing the macro features, we see that the correlation feature in the second-order statistics provides a measure of the linearity in the directions on an object's thermal image. The wood wall presents the highest correlation value amongst the extended objects as a result of its vertical boards and wood grains on the surface. Though the bark on the pine tree log tends to extend in a vertical direction, the zigzag design results in a lack of linear structure and the lowest correlation value amongst the compact objects. Similar to the uniformity in the first-order statistical feature, energy in the second-order case measures the intensity of gray-level (radiant) transitions in the thermal image of an object. Values for both uniformity and energy increase as the gray-level becomes more uniformly distributed and are unity for a thermal image of an object with a constant surface radiance. For the extended objects, the brick wall shows the highest uniformity and energy values since it displays less intense radiant transitions compared to the hedges and wood wall. The steel pole presents the highest uniformity and energy feature values for the compact objects due to its relatively constant surface radiance. Since homogeneity is similar to energy, its results are consistent with those presented by the energy feature. Contrary to uniformity, energy, and homogeneity tending to increase in value for objects with a uniform or constant surface radiance, contrast and entropy feature values increase for objects with more variations (randomness or complexity) in radiant emissions. The wood wall presents a higher contrast feature value for both the first- and second-order statistical cases compared to the brick wall and hedges. The larger amount of variation in the radiance for the wood wall is contributed by both the radiant patterns of wood grains on the surface of the boards and the surface radiances in the foreground emitted through the gaps of the wood boards. For the compact objects, the pine tree log displays the highest contrast in both the first- and second-order statistic cases as a result of the large variations in the radiance from the bark pattern. Entropy is a measure of complexity (or randomness) in an object's thermal image. Since the entropy feature tends to be sensitive to the variations in the radiance of an objects thermal image, its results are consistent with the contrast feature. For the extended objects, the hedges have a high entropy value for both the first- and second-order statistics as expected. However, the wood wall presents the highest entropy values due to the feature's sensitivity to the combined effects of varying radiation emitted from the wood grains on the surface of the boards and the surface radiances in the foreground emitted through the gaps of the wood boards. The rough surface and the zigzag pattern of the bark on the pine tree log results in a more complex surface compared to the concrete cylinder and steel pole. Therefore, the pine tree log has the highest entropy amongst the compact objects. The concrete cylinder has the second highest entropy due to its mixture of stones and cement creating a random radiant pattern compared to the steel pole's smooth radiant surface.

As we see, the micro and macro features all generate unique representations of a non-heat generating object from the given object's thermal image. The meteorological features serve to estimate the current and historical effects of the diurnal cycle of solar energy on the amount of radiance emitted from an object's surface. Consequently, not only will the micro and macro features provide inter-class variation to distinguish one object class from another, these features will also display intra-class variations due to the variations of the meteorological features. Our performance and feature selection process presented in Chap. 4 will prove that the most favorable feature sets are those that contain contributions from all the feature types - meteorological, micro, and macro.

3.8 Curvature Algorithm

In Sect. 3.5 we discussed the factors that cause variations in radiance on cylindrical objects. These factors consisting of directional variation of emissivity, irradiance from sources in the background, and/or halo effect can also assist us in deriving a curvature algorithm used to distinguish compact objects from extended objects. Our curvature algorithm is presented in Table 3.5. In Step 1, the algorithm computes the average of radiances at the center, vertical, horizontal, and diagonal segments of the object's thermal image. In Step 2, the absolute differences between the average radiance at the center and the average radiance at the neighboring vertical,

Table 3.5 Curvature Algorithm used to distinguish compact and extended objects.

Curvature Algorithm
Step 1: Compute the average radiance of an object's thermal image at center ($\bar{R}_c$), verticals ($\bar{R}_{v1}, \bar{R}_{v2}$), horizontals ($\bar{R}_{h1}, \bar{R}_{h2}$), and diagonals ($\bar{R}_{d1}, \bar{R}_{d2}, \bar{R}_{d3}, \bar{R}_{d4}$) as displayed in the diagram to the right. 
Step 2: Compute the absolute radiance differences: $C_{v1} = \lvert \bar{R}_c - \bar{R}_{v1} \rvert, C_{v2} = \lvert \bar{R}_c - \bar{R}_{v2} \rvert, C_{h1} = \lvert \bar{R}_c - \bar{R}_{h1} \rvert, C_{h2} = \lvert \bar{R}_c - \bar{R}_{h2} \rvert,$ $C_{d1} = \lvert \bar{R}_c - \bar{R}_{d1} \rvert, C_{d2} = \lvert \bar{R}_c - \bar{R}_{d2} \rvert, C_{d3} = \lvert \bar{R}_c - \bar{R}_{d3} \rvert, C_{d4} = \lvert \bar{R}_c - \bar{R}_{d4} \rvert$
Step 3: For a given threshold value C_T, the following classifications are concluded: If $$[(C_{v1} \wedge C_{v2}) < C_T \wedge (C_{h1} \wedge C_{h2}) \geq C_T] \vee [(C_{h1} \wedge C_{h2}) < C_T \wedge (C_{v1} \wedge C_{v2}) \geq C_T]$$ $$\vee [(C_{d3} \wedge C_{d4}) < C_T \wedge (C_{d1} \wedge C_{d2}) \geq C_T] \vee [(C_{d1} \wedge C_{d2}) < C_T \wedge (C_{d3} \wedge C_{d4}) \geq C_T],$$ then the object is classified as compact-cylindrical. ElseIf $$(C_{v1} \wedge C_{v2} \wedge C_{h1} \wedge C_{h2} \wedge C_{d1} \wedge C_{d2} \wedge C_{d3} \wedge C_{d4}) > C_T,$$ then the object is classified as compact-spherical. ElseIf at least one pair of image segments symmetric about the center segment have absolute radiance difference values (from Step 2) of at least that of the given threshold value C_T, then the object is classified as compact (without regard to being cylindrical or spherical). Else the object is classified as extended.

horizontal, and diagonal segments are computed. The absolute difference is chosen since the periphery of an object could have a smaller gray-level value than the center or vice versa, depending on the effects of the directional variation of emissivity, irradiance from sources in the background, and/or halo effect. In Step 3, these absolute differences are compared to a given threshold value C_T and conclude whether an object is compact-cylindrical, compact-spherical, compact (without regards to being cylindrical or spherical in shape) or extended. The rule for a compact-cylindrical object in Step 3 takes into consideration the possibility of a cylindrical object tilted at different orientations. We can also identify compact objects that display minimal directional variation of emissivity but still present variations in radiance from the center to the peripheries due to background irradiance and/or the halo effect. The square steel pole displayed in Fig. 3.18 (b) is an example of this type of compact object. Since these objects are not cylindrical or spherical, we label them as compact (without regards to being cylindrical or spherical in shape).

Fig. 3.18 Visible and thermal images of objects used to demonstrate curvature algorithm. Segmented regions in thermal images are used to compute the average radiances used in the curvature algorithm. (a) tree, (b) square metal pole, (c) brick wall.

As a demonstration of the curvature algorithm, consider the tree, square metal pole, and brick wall in Fig. 3.18. The segmented regions in thermal images are used to compute the average radiances $\overline{R}$ used in the curvature algorithm. The results of the computations from the curvature algorithm are presented in Table 3.6. With a threshold value of $C_T = 1.1$, the tree would be assigned as a com-

Table 3.6 Curvature Algorithm demonstration results using objects in Fig. 3.17.

Object	C_{v1}	C_{v2}	C_{h1}	C_{h2}	C_{d1}	C_{d2}	C_{d3}	C_{d4}
Tree	1.0061	0.4933	3.6701	8.4768	3.8899	10.3757	3.7682	3.818
Square Metal Pole	0.3926	0.0234	4.9808	0.8537	5.963	1.187	0.2398	6.9402
Brick Wall	0.0814	1.5147	2.2482	0.0028	0.3565	1.2591	0.9742	0.491

pact-cylindrical object, square metal pole as a compact object (without regards to being cylindrical or spherical in shape), and the brick wall would be assigned as an extended object. As we will also mention in Chap. 6, with further investigation the curvature algorithm has potential to serve as an exceptional technique to distinguish compact objects from extended objects.

3.9 Summary

In this chapter we discussed the thermal features used in our research to classify non-heat generating objects. Examples were provided to illustrate the value of our features in distinguishing non-heat generating objects. A summary of our equations for these thermal features is displayed in Table 3.3. By generating feature values from the thermal images of non-heat generating objects, we have seen how interpreting the effects of the outdoor environment and thermal properties of objects on their feature values is a subtle process. We also presented a curvature algorithm to assist us in distinguishing compact objects from extended objects. In the next chapter we will select the most favorable sets from these features based on their performance with various classifiers. We will also analyze the behavior of our most favorable set of features with variations in the viewing angle with the target, thermal image window size, and rotational orientation of the target.

References

[1] Duda RO, Hart PE et al (2001) Pattern Classification. 2nd edn. Wiley, New York
[2] Gonzalez RC, Woods RE et al (2004) Digital Image Processing using MATLAB. Pearson/Prentice Hall, Upper Saddle River, NJ
[3] Theodoridis S, Koutroumbas K (2006) Pattern Recognition. 3rd edn. Academic Press, San Diego, CA
[4] Tomita F, Tsuji S (1990) Computer Analysis of Visual Textures. Kluwer Academic Publishers, Boston
[5] Tuceryan M, Jain AK (1999) Texture Analysis. In: Chen CH, Pau L- et al (ed) Handbook of Pattern Recognition & Computer Vision. 2nd edn. World Scientific, River Edge, NJ
[6] Wood J (1996) Invariant Pattern Recognition: A Review. Pattern Recognition 29(1):1–17
[7] Zhang JG, Tan TN (2002) Brief Review of Invariant Texture Analysis Methods. Pattern Recognition 35(3):735–747

[8] Hu M (1962) Visual Pattern Recognition by Moment Invariants. IRE Transactions on Information Theory 8(2):179–187
[9] Reiss TH (1991) The Revised Fundamental Theorem of Moment Invariants. IEEE Transactions on Pattern Analysis and Machine Intelligence 13(8):830–834
[10] Flusser J (2000) On the Independence of Rotation Moment Invariants. Pattern Recognition 33(9):1405–1410
[11] Flusser J (2006) Moment Invariants in Image Analysis. Transactions on Engineering, Computing and Technology VII:196–201
[12] Suk T, Flusser J (2003) Combined Blur and Affine Moment Invariants and their use in Pattern Recognition. Pattern Recognition 36(12):2895–2907
[13] Van Gool L, Moons T et al (1996) Affine/Photometric invariants for planar intensity patterns. Computer Vision, ECCV '96: 4th European Conference on Computer Vision-Springer, :642–651
[14] Wang L, Healey G (1998) Using Zernike Moments for the Illumination and Geometry Invariant Classification of Multispectral Texture. IEEE Transactions on Image Processing 7(2):196–203
[15] Grace AE, Spann M (1991) A Comparison between Fourier-Mellin Descriptors and Moment Based Features for Invariant Object Recognition using Neural Networks. Pattern Recognition Letters 12(10):635–643
[16] Casasent D, Psaltis D (1976) Scale Invariant Optical Transform. Optical Engineering 15(3):258–261
[17] Casasent D, Psaltis D (1976) Position, Rotation, and Scale Invariant Optical Correlation. Applied Optics 15(7):1795–1799
[18] Sheng Y, Arsenault HH (1986) Experiments on Pattern Recognition using Invariant Fourier-Mellin Descriptors. Journal of the Optical Society of America A 3(6):771–776
[19] Sheng Y, Lejeune C et al (1988) Frequency-Domain Fourier-Mellin Descriptors for Invariant Pattern Recognition. Optical Engineering 27(5):354–357
[20] Sheng Y, Duvernoy J (1986) Circular-Fourier-Radial-Mellin Transform Descriptors for Pattern Recognition. Journal of the Optical Society of America A 3(6):885–888
[21] Treptow A, Cielniak G et al (2006) Real-Time People Tracking for Mobile Robots using Thermal Vision. Robotics and Autonomous Systems 54(9):729–739
[22] Prokoski F (2000) History, Current Status, and Future of Infrared Identification. Proceedings of IEEE Workshop on Computer Vision Beyond the Visible Spectrum: Methods and Applications :5–14
[23] Lu YJ, Hsu YH et al (1992) Vehicle Classification using Infrared Image-Analysis. Journal of Transportation Engineering-Asce 118(2):223–240
[24] Sevigny L, Hvedstrup-Jensen G et al (1983) Discrimination and Classification of Vehicles in Natural Scenes from Thermal Imagery. Computer Vision, Graphics, and Image Processing, 24(2):229–243
[25] Fang YC, Wu BW (2007) Neural Network Application for Thermal Image Recognition of Low-Resolution Objects. Journal of Optics A-Pure and Applied Optics 9(2):134–144
[26] Strickland RN (1994) Infrared Techniques for Military Applications. In: Maldague XPV (ed) Infrared Methodology and Technology. Gordon and Breach Science Publishers, Yverdon, Switzerland ; Langhorne, Pa., U.S.A.
[27] Nandhakumar N, Aggarwal JK (1988) Integrated Analysis of Thermal and Visual Images for Scene Interpretation. IEEE Transactions on Pattern Analysis and Machine Intelligence 10(4):469–481
[28] Michel JD, Nandhakumar N et al (1998) Geometric, Algebraic, and Thermophysical Techniques for Object Recognition in IR Imagery. Computer Vision and Image Understanding 72(1):84–97
[29] Bharadwaj P, Carin L (2002) Infrared-Image Classification using Hidden Markov Trees. IEEE Transactions on Pattern Analysis and Machine Intelligence 24(10):1394–1398

[30] El Maadi A, Maldague X (2007) Outdoor Infrared Video Surveillance: A Novel Dynamic Technique for the Subtraction of a Changing Background of IR Images. Infrared Physics & Technology 49(3):261–265
[31] Horney T, Ahlberg J et al (APR 2004) An information system for target recognition. Proceedings of SPIE, Multisensor, Multisource Information Fusion: Architectures, Algorithms, and Applications 2004:163–175
[32] Maldague XPV (2001) Theory and Practice of Infrared Technology for Nondestructive Testing. Wiley, New York
[33] Meriaudeau F (February 2007) Infrared imaging and machine vision. Proceedings of SPIE, Machine Vision Applications in Industrial Inspection XV:650308-1-650308-11
[34] Maldague X (1994) Infrared Methodology and Technology. Gordon and Breach Science Publishers, Yverdon, Switzerland ; Langhorne, Pa., U.S.A.
[35] Al-Habaibeh A, Shi F et al (2004) A Novel Approach for Quality Control System using Sensor Fusion of Infrared and Visual Image Processing for Laser Sealing of Food Containers. Measurement Science & Technology 15(10):1995–2000
[36] Ginesu G, Giusto DD et al (2004) Detection of Foreign Bodies in Food by Thermal Image Processing. IEEE Transactions on Industrial Electronics 51(2):480–490
[37] Madding RP (APR 1999) Emissivity measurement and temperature correction accuracy considerations. SPIE Conference on Thermosense XXI
[38] Holst GC (2000) Common Sense Approach to Thermal Imaging. JCD Pub.; co-published by SPIE Optical Engineering Press, Winter Park, Fla.; Bellingham, Wash.
[39] Cuenca J, Sobrino JA (2004) Experimental Measurements for Studying Angular and Spectral Variation of Thermal Infrared Emissivity. Applied Optics 43(23):4598–4602
[40] Sobrino JA, Cuenca J (1999) Angular Variation of Thermal Infrared Emissivity for some Natural Surfaces from Experimental Measurements. Applied Optics 38(18):3931–3936
[41] French AN, Schmugge TJ et al (2000) Discrimination of Senescent Vegetation using Thermal Emissivity Contrast. Remote Sensing of Environment 74(2):249–254
[42] Weszka JS, Dyer CR et al (1976) A Comparitive Study of Texture Measures for Terrain Classification. IEEE Transactions on Systems, Man, and Cybernetics SMC-6(4):269–285
[43] Tamura H, Mori S et al (1978) Textural Features Corresponding to Visual Perception. IEEE Transactions on Systems, Man and Cybernetics SMC-8(6):460–473
[44] Beaudoin JL, Bissieux C (1994) Theoretical Aspects of Infrared Radiation. In: Maldague X (ed) Infrared Methodology and Technology, Nondestructive Testing Monographs and Tracts. Gordon and Breach Science Publishers, Yverdon, Switzerland ; Langhorne, Pa., U.S.A.
[45] Incropera FP, Dewitt DP et al (2007) Fundamentals of Heat and Mass Transfer. 6th edn. John Wiley, Hoboken, NJ
[46] Modest MF (2003) Radiative Heat Transfer. 2nd edn. Academic Press, Amsterdam ; Boston
[47] Shannon CE (1948) A Mathematical Theory of Communication. The Bell System Technical Journal 27:379–423, 623–656
[48] Tribus M, McIrvine EC (1971) Energy and Information. Scientific American 224(3): 179–188
[49] Haralick RM, Shanmugam K et al (1973) Textural Features for Image Classification. IEEE Transactions on Systems, Man, and Cybernetics SMC-3(6):610–621
[50] Siew LH, Hodgson RM et al (1988) Texture Measures for Carpet Wear Assessment. IEEE Transactions on Pattern Analysis and Machine Intelligence 10(1):92–105
[51] MathWorks I (2006) Image Processing Toolbox : For use with MATLAB® : User's Guide. Version 5 edn. MathWorks, Inc., Natick, MA

4 Thermal Feature Selection

Abstract In this chapter, we evaluate the performance of various classification models to identify the most favorable feature vectors for our extended and compact objects. We will show that there is no single "optimal" feature vector but a set of "most favorable" feature vectors associated with various classifiers for both the extend and compact object classes. Moreover, the most favorable feature vectors are those that contain contributions from all the feature types – meteorological, micro, and macro.

4.1 Introduction

In the previous chapter, we generated 21 thermal features from three categories – meteorological, micro, and macro. This chapter will present the third step in our pattern classification model design process – thermal feature selection. In the current and subsequent chapters, we will assume that the robotic thermal imaging system has already used algorithms to detect the presence of an unknown non-heat generating object, identified the object as being either extended or compact, and segmented the object to generate our thermal features. In the context of this research, we have defined non-heat generating objects as objects that are not a source for their own emission of thermal energy, and so exclude people, animals, vehicles, etc. The extended objects consist of objects that extend laterally beyond the thermal camera's lateral field of view, such as brick walls, hedges, picket fences, and wood walls. The compact objects consist of objects that are completely within the thermal camera's lateral field of view, such as steel poles and trees. Our analysis in the classification model design process will consider the extended and compact categories separately as disjoint sets of object classes.

The current goal is to select sets of features from the three feature categories (meteorological, micro, and macro) that provide the most favorable information to allow us to classify the unknown non-heat generating object with minimal error.

Each of these sets of features is called a feature vector (or pattern). We will begin our feature selection process with a preliminary feature analysis to explore for any outliers in the data and eliminate redundant features while avoiding any "data dredging" and retaining only those sets of features that enhance our ability to distinguish object classes. Since the performance of a classifier is a function of the feature vector, the subsequent evaluation of classifiers and selection of feature sets are done simultaneously. Our selection process will involve an exhaustive search using a high performance computing system to analyze the classification performance of over 290,000 feature combinations spanning up to 18 dimensions. Common in the assessment of all feature vector candidates is their ability to minimize the error in classifying non-heat generating objects. We will see that there is no single "optimal" feature vector but we will have a set of "most favorable" feature vectors associated with various classifiers. Moreover, our process will prove that the most favorable feature vectors are those that contain contributions from all the feature types – meteorological, micro, and macro.

4.2 "No Free Lunch" Classifiers

Selecting the most favorable sets of feature vectors is not a trivial process. Each feature vector is selected based on its performance with a given classifier. Therefore, the feature vector and classifier combination that results in minimum classification errors becomes the most favorable pattern classification model. However, as we discussed in Chap. 3, there is no universal feature vector according to the Ugly Duckling Theorem. Similarly, according to the No Free Lunch Theorem [1], there is no universal classifier or learning algorithm. The classifier is chosen based on how well it performs for a specific pattern classification application. Since the performance of a classifier is a function of a feature vector, there is obviously no universal pattern classification model. Our application makes choosing a pattern classification model even more complex due to the variations in the thermal feature values caused by the diurnal cycle of solar energy. We will see in subsequent chapters that each of our object classes will have their own set (or committee) of most favorable pattern classification models. Each committee will result in the most favorable performance on unknown patterns from their respective object class, but may not perform well on patterns from other object classes. The combination of these committees will result in a model that exploits the complementary information found in each classification model and improves overall performance.

There are many choices for the type of classifier or learning algorithm to use in pattern classification models. Reviews of pattern recognition methods and theory are found in [1]–[14]. The most popular approaches for pattern recognition are statistical classification, template matching, and neural networks. The method of choice usually is based objectively on which approach results in minimum classification errors and/or subjectively on which approach provides the operator with the desired data format in the output. Our desired approach is the one that results in minimum classification errors while retaining the original physical interpretation of the information in the signal data throughout the entire classification process. We choose not to use neural networks since this approach tends to conceal the original physical interpretation and statistics of the data [2]. In template matching, an unknown pattern is compared with a library of templates (or prototypes). A similarity (or correlation) measure is used to decide which of these templates the unknown pattern matches best. One possibility for creating a template is by computing a mean reference pattern from an object class's training set. A major disadvantage of template matching is that it tends to fail with large intra-class variations among the patterns [2]. Consequently, template matching is not an appropriate method for our application since our thermal features experience intra-class variations due to the diurnal cycle of solar energy. In statistical classification, each object class is represented by a distribution of feature vectors that are chosen to maximize the distinction between each object class. The goal is to assign an unknown pattern to one of the object classes by considering the combination of these distributions of feature vectors and any prior knowledge regarding each object class. This approach affords the ability to classify unknown patterns from distributions that display intra-class variations. In our case, these variations of the feature vectors within each object class are due to the diurnal cycle of solar energy. Moreover, the statistical classification approach retains the original physical interpretation of the information in the signal data throughout the entire classification process. Consequently, statistical classification seems to be the most favorable method for our application.

Statistical classification is further divided into two categories – supervised and unsupervised. In unsupervised classification, class labeling of the data is not available and techniques such as clustering are used to identify features that assist in distinguishing groups. Once the structure of the data is understood, an unknown pattern can be assigned to one of the groups. As introduced in Chap. 2, our application consists of labeled object classes – brick walls, hedges, wood picket fences, wood walls, steel poles, and trees. Consequently, our application is categorized as supervised classification where learning involves labeled classes of data. In our case, an unknown pattern is assigned to one of our predefined object classes.

For a given pattern classification application the density function representing the distribution of the data in each object class is known or unknown. Cases where the density function is known are called parametric techniques. For example, a parametric technique could use a Gaussian density estimation for an object class with a known normal distribution. Due to the variations in the thermal feature values caused by the diurnal cycle of solar energy, we will not assume a formal density function for the distribution of the data in each object class. Therefore, we will make use of nonparametric techniques for our pattern classification application.

Two popular approaches for nonparametric techniques are the decision boundary approach and the probabilistic approach. The decision boundary approach for nonparametric cases involves the design of a discriminant function that defines the decision boundaries used to distinguish one object class from another. However, these discriminant functions tend to disguise probabilistic information in the data and the original physical interpretation of the information in the signal data though transformations with weight vectors. On the other hand, the probabilistic approach assigns an unknown pattern to one of the object classes based on a decision rule derived from posterior probabilities that consider the combination of density function estimations for the distributions of the data and any prior knowledge regarding each object class. The probabilistic approach is our choice for a nonparametric technique. In summary, our approach is an application of statistical pattern classification where learning involves labeled classes of data (supervised classification), assumes no formal structure regarding the density of the data in the classes (nonparametric density estimation), and makes direct use of posterior probabilities when making decisions regarding class assignments (probabilistic approach).

The remainder of this chapter will proceed as follows. In Sect. 4.3, we will present a preliminary feature analysis to assess the quality of our training data and eliminate any redundant features. Section 4.4 will present the nonparametric classifiers that we will use during the feature selection process. In Sect. 4.5, we will discuss and implement performance criteria and feature selection methods to select our most favorable features for extended and compact objects. In Sect. 4.6, we will perform a sensitivity analysis to explore the effects of variations in the camera's viewing angle, window size of the thermal scene, and rotational orientation of the target on the feature values and classification performance of a classifier involving selected feature vectors from our most favorable sets. We will provide a summary of the chapter in Sect. 4.7. The methods presented in this chapter were implemented with assistance by a Matlab toolbox for pattern recognition known as PRTools4 [15].

4.3 Preliminary Feature Analysis

In this section, we perform a preliminary feature analysis (exploratory data analysis or initial data analysis) of the thermal features (see Chap. 3) generated from our training data (see Sect. 2.3). Since the training data has a direct effect on the learning process of the pattern classification model, assessing the quality of the data is a crucial step. Our preliminary feature analysis consists of three steps. First, we will analyze the data for any outliers. Second, we will standardize the data values for each thermal feature. Each set of thermal feature values in the training data were standardized over all object classes within each of the extended and compact object categories. We used the following standardization equation presented in a study of standardization methods by Milligan and Cooper [16]:

$$Z5 = \frac{(X - Min(X))}{(Max(X) - Min(X))} \quad (4.1)$$

where X is the original thermal feature value being standardized. The Z5 standardization method is bounded by 0.0 and 1.0 with at least one feature value at each of the end points. We adopted the Z5 standardization method since it does not require an assumed formal density function for the distribution of the data. Furthermore, Milligan and Cooper's study showed that methods such as Z5 involving the range of the data values as the divisor offer the best recovery of the underlying data structure. Third, we will use scatter plots for an initial feature redundancy reduction. The goal in feature redundancy reduction is to retain features where the relationship between pairs of features improves class separation and eliminate features where strong linear relationships result in redundancy. Since preliminary feature analysis is a very subjective process, we will avoid any data dredging [17] that could result in an over-fitted pattern classification model and/or reducing the quality of our representative data set.

We did not identify any outliers among the 424 extended objects in the training data set presented in Table 2.1. After standardizing the feature values using Eq. 4.1, we used scatter plots to study the relationship between each pair of thermal features found in Table 3.3. Since *Co*1 has a strong relation with *So*1 as shown in Fig. 4.1, *Co*1 is eliminated from the choice of thermal features. Similarly, *Uo*1 is removed due to its strong relation with *En*1 as displayed in Fig. 4.2. Additionally, *To*1 was eliminated since the majority of its vales are between 0.61 and 0.63 as presented in Fig. 4.3. This small interval containing the majority of the *To*1 feature values results in a lack of separation between the object classes. The scatter matrix for the remaining 18 extended object thermal features is presented in Fig. 4.4. We still observe other pairs of thermal features with strong relationships; however, we will retain these features for further analysis when we assess the performance of the feature combinations with various classifiers. For example, as we see in Fig. 4.4, both *Mor*1 and *Mob*1 display a strong relation with *Lr* and *Lb*, respectively, due to the dependencies found in their thermal feature

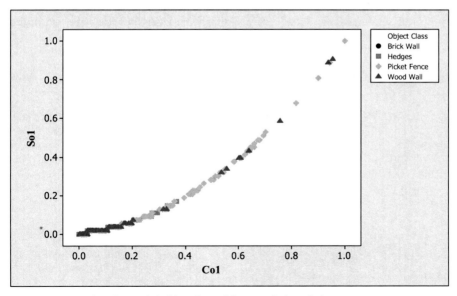

Fig. 4.1 Scatter plot of extended object thermal features Co1 vs. So1.

100   4 Thermal Feature Selection

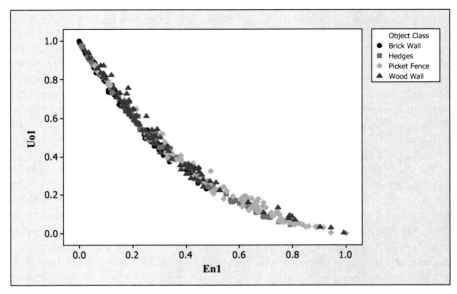

Fig. 4.2 Scatter plot of extended object thermal features Uo1 vs. En1.

equations (see Table 3.3). As expected, we also see a strong relationship between $En1$ and $En2$. As discussed in our application of the thermal features in Sect. 3.7, the characteristics of $Ho2$ are similar to $Er2$. As a result, $Ho2$ displays a strong relationship with $Er2$ in the scatter matrix. We also noted that contrary to $Er2$ and $Ho2$ increasing in value for objects with a uniform or constant surface radiance, $Co2$, $En1$, and $En2$ increase in value for objects with more variations (or complexity) in radiant emissions. Consequently, these characteristics are observed in the strong relationship of data trends with negative slopes.

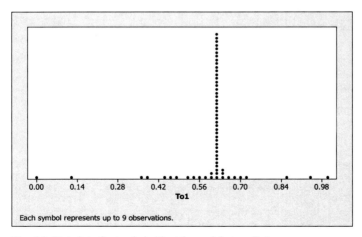

Fig. 4.3 Dot plot of extended object thermal feature To1.

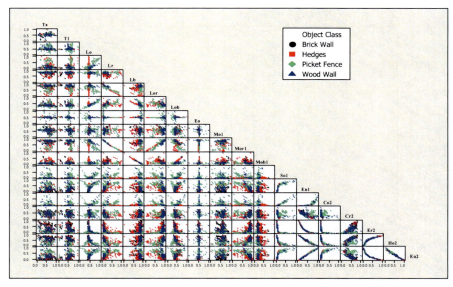

Fig. 4.4 Scatter matrix of remaining extended object thermal features after a preliminary feature analysis.

For the compact objects, two thermal feature values for emissivity were identified as outliers. As we discussed in Chap. 3, the emissivity values computed by Eq. 3.11 could be quite sensitive to variations in the thermal radiance of the object, reference emitter, and aluminum foil. For instance, as the radiance of the reference emitter and the aluminum foil approach the same value, the denominator in the equation for emissivity will become very small (either positive or negative). As a result, the value of the emissivity in Eq. 3.11 would take on very large values (either positive or negative). This is the situation for our two outliers. One of the thermal images of a steel pole had $Lo = 95.2844$, $Lr = 95.0581$, and $Lb = 95.0479$. The computed emissivity of $Eo = 22.4907$ was identified as an outlier. The other outlier involved the thermal image of a tree with $Lo = 94.6489$, $Lr = 94.0923$, and $Lb = 94.1144$. In this case, the computed emissivity was $Eo = -23.46$. The thermal image of the steel pole was captured at 1049 hrs on 21 March 2007 with an ambient temperature of $45.6°F$ and cloud coverage at a high ceiling altitude. The thermal image of the tree was captured at 1738 hrs on 25 March 2007 with an ambient temperature of $51.4°F$ and no cloud coverage. Thus, we can conclude the environmental conditions and viewing angle of the thermal camera were just right for the target and surrounding surfaces to have approximately the same level of thermal radiant emissions. This phenomenon, known as thermal crossover [18], resulted in the minimal thermal contrast between the surfaces of objects and the surrounding that caused the extreme emissivity values. Consequently, the thermal images of the steel pole and tree that created these emissivity outliers were removed from the training data set. Table 2.1 displays the remaining 636 compact objects used in the training data set.

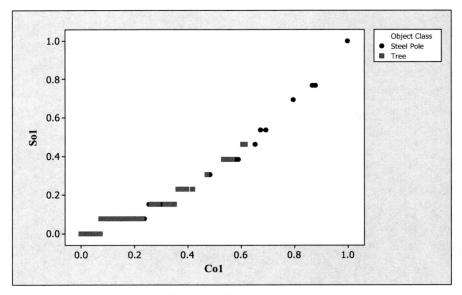

Fig. 4.5 Scatter plot of compact object thermal features $Co1$ vs. $So1$.

In Chap. 6, we will discuss how these periods of thermal crossover could result in a limitation to our ability to classify non-heat generating objects in an outdoor environment using a thermal infrared imaging sensor. We will also present a method that incorporates a thermal contrast threshold rule into the detection phase of the classification process that requires a minimum amount of contrast in the scene to use the thermal infrared imaging sensor. If the rule is not satisfied, the autonomous robot must reject the use of the thermal imaging sensor and rely on other sensors such as ultrasound to assist in classifying the target.

After standardizing the feature values of the compact objects using Eq. 4.1, we used scatter plots to study the relationship between each pair of thermal features found in Table 3.3. As noted in Chap. 3, the thermal features $Mo1$, $Mor1$, and $Mob1$ will not apply to the compact objects since $Lo = Mo1$. Since $Co1$ has a strong relation with $So1$ as shown in Fig. 4.5, $So1$ is eliminated from the choice of thermal features. The feature $So1$ is eliminated since $Co1$ consists of more distinct feature values than $So1$ as displayed in Fig. 4.5. The thermal feature $Uo1$ is removed due to its strong relation with $En1$ as displayed in Fig. 4.6. Additionally, $To1$ was eliminated since the majority of its values are between 0.21 and 0.23 as presented in Fig. 4.7. As with the extended object case, this small interval containing the majority of the $To1$ feature values results in a lack of separation between the object classes. The scatter matrix for the remaining 15 compact object thermal features is presented in Fig. 4.8. The remaining pairs of thermal features with strong relationships in the scatter matrix will be retained for further analysis when we assess the performance of the feature combinations with vari-

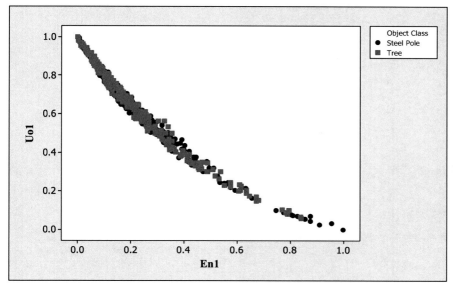

Fig. 4.6 Scatter plot of compact object thermal features Uo1 vs. En1.

ous classifiers. As expected, *Co*1 and *En*1 display similar characteristics by their strong relationship with a positive sloping trend in data. Furthermore, *Er*2 and *En*2 present opposing attributes by displaying a negative sloping trend in their data. Interestingly, we see strong relationships within each object class involving the pairs of features (*En*1, *En*2), (*Co*1, *En*2), (*Co*2, *En*1), (*Co*2, *En*2), and (*Er*2, *Ho*2) that result in an increasing separation between the two object classes' data from a common origin.

A reoccurring observation in the scatter plots for both the extended and compact objects is that the data for each object class tends to diverge from the other object classes beginning at a common origin. We see a separation in the object classes that is dependent on the variation in thermal features due to the diurnal cycle of solar energy. Consequently, the origins represent thermal conditions in the environment that are just right for the feature values to not display much distinction between object classes. In typical classification applications involving controlled environments, the feature values for each object class tend to form compact hyperspherical or hyperellipsoidal clusters with no common origin amongst the object classes. These applications normally use traditional metrics to choose a set of features for the classification model such as the inter/intra class distance where the most favorable set of features is the one that results in a large distance between object class clusters (interclass) and small distance between feature vectors within each object class (intraclass). Since our application involves a dynamic outdoor environment, we are dealing with a more complex situation that requires non-traditional methods.

104   4 Thermal Feature Selection

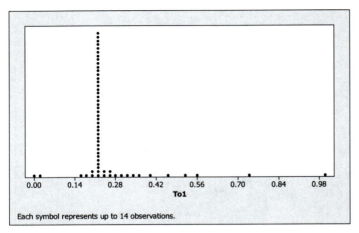

Fig. 4.7 Dot plot of compact object thermal feature To1.

Due to the complexity of classification applications involving outdoor images, we have only found three relevant attempts in the literature to classify features generated from the images of outdoor objects that vary with the availability of solar radiation. Buluswar and Draper present a color-based recognition application under varying illumination in an outdoor environment using features based on RGB (Red, Green, Blue) space to classify the color of surfaces for autonomous vehicles [19]

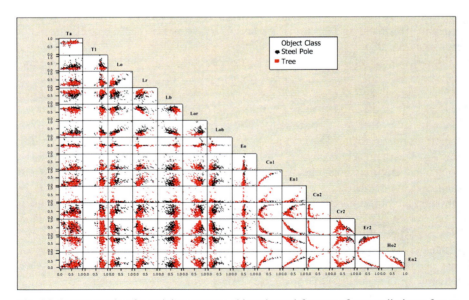

Fig. 4.8 Scatter matrix of remaining compact object thermal features after a preliminary feature analysis.

and machine vision [20]. A representative training data set consisting of color features generated from images of natural objects in an outdoor environment covering a wide range of illumination conditions is used in a maximum likelihood classifier in [21]. The classification using the color stereo camera is complemented by a single axis LADAR sensor for autonomous navigation in cross-country environments.

As we will see later, the diverging nature of the structure in our object classes' clusters will continue in higher dimensions. Since our object classes' clusters resemble a conical structure, they will be called hyperconoidal clusters. These hyperconoidal clusters are the cornerstones of our research that inspired our novel method for classifying non-heat generating objects in an outdoor environment that we will present in Chap. 5.

4.4 Classifiers

In this section we discuss our nonparametric classifiers that will have a probabilistic approach when making decisions regarding class assignments. The three classifiers used in our feature selection process are Bayesian, K-Nearest-Neighbor (KNN), and Parzen.

4.4.1 Bayesian Classifier

In this section we will derive our Bayesian classifier that uses a KNN density estimation. Suppose we want to find the probability of an arbitrary object class O_j, $j = 1,...,J$, being present given that we generated a feature vector $\underline{D}_n$ from the signal emitted by the object and received by our sensor n. In mathematical terms, we seek to find the conditional probability $P(O_j | \underline{D}_n)$. Intuitively, we would think that this condition somehow depends on the joint probability $P(\underline{D}_n, O_j)$ that we obtained the feature vector $\underline{D}_n$ from the signal and it belongs to the object class O_j. Our joint probability is defined using the product rule $P(\underline{D}_n, O_j) = P(\underline{D}_n | O_j)P(O_j)$ where the conditional probability $P(\underline{D}_n | O_j)$ provides a measure of the chance that we would have obtained the values in the feature vector $\underline{D}_n$ if the object class O_j was given to be present and $P(O_j)$ provides a measure of our state of knowledge regarding the object class being present before any signal data is collected using the sensor. Since the probability of both $\underline{D}_n$ and O_j being true must be logically equivalent to O_j and $\underline{D}_n$ being true so that $P(\underline{D}_n, O_j) = P(O_j, \underline{D}_n)$, we must have $P(\underline{D}_n | O_j)P(O_j) = P(O_j | \underline{D}_n)P(\underline{D}_n)$.

Since all the joint probabilities $P(\underline{D}_n, O_j)$, for $j = 1...J$, are mutually exclusive, the unconditional probability

$$P(\underline{D}_n) = P(\underline{D}_n, O_1) + P(\underline{D}_n, O_2) + ... + P(\underline{D}_n, O_J)$$
$$= \sum_{j=1}^{J} P(\underline{D}_n | O_j) P(O_j) \qquad (4.2)$$

is the total probability of obtaining the feature vector $\underline{D}_n$, irrespective of object class membership. Thus, we have

$$P(O_j | \underline{D}_n) = \frac{P(\underline{D}_n | O_j) P(O_j)}{P(\underline{D}_n)}. \qquad (4.3)$$

This expression is known as Bayes' theorem (or Bayes' formula), named after Reverend Thomas Bayes (1702 – 1761). The quantity $P(O_j | \underline{D}_n)$ is called the posterior probability since it gives the probability of the object class being O_j after obtaining the measured feature vector $\underline{D}_n$. The quantity $P(\underline{D}_n | O_j)$ is called the likelihood function since it provides a measure of the chance that we would have obtained the values in the feature vector $\underline{D}_n$ if the object class O_j was given to be present. As noted by R. A. Fisher [22], though the likelihood function is provided in the form of a conditional probability, it is not necessarily a probability density function since the integral of a likelihood function may not equal to one. Consequently, we will refer to the likelihood function as a probability density estimation. The quantity $P(O_j)$ is called the prior probability since it represents our state of knowledge regarding the object class being present before any signal data is collected using the sensor. For example, if we feel that all the object classes could exist in our robot's local area of operation or have no reason to believe that one object class is more likely to be identified over another, then the "principle of indifference" prevails and we assign equal priors for all the object classes. The quantity $P(\underline{D}_n)$ is a normalization parameter (known as the evidence) that ensures that the posterior probabilities sum to unity.

From Bayes' theorem we can form Bayes' decision rule that allows us to minimize the probability of misclassification by selecting the object class O_k having the largest posterior probability compared to posterior probabilities of the other object classes. That is, given a feature vector $\underline{D}_n$ obtained from the signal received by our sensor, we conclude that the source of the signal is object class O_k if

$$P(O_k | \underline{D}_n) > P(O_l | \underline{D}_n) \qquad (4.4)$$

for all $k \neq l$.

As noted earlier, the likelihood function provides a measure of the chance that we would have obtained the values in a feature vector if an object class was given to be present. Otherwise, the likelihood function is a probability density estimation in the data space. Formally, a probability density function $p(x)$ is used to find the probability that a variable x lies within an interval from $x = a$ to $x = b$ and is given by

$$P(x \in [a, b]) = \int_a^b p(x)\, dx. \tag{4.5}$$

If the density function is known based on the distribution of the data and we do not expect the distribution to vary, then we could choose parametric techniques to formulate our probability density function. However, if we expect our data to vary based on environmental factors and our actual density function is unknown, we should seek nonparametric methods that can be used with arbitrary distributions to derive our probability density estimation.

The general method in formulating an estimate for an unknown probability density function $p(x)$ is discussed by Duda, Hart, and Stork [1] and Bishop [4] as follows. Suppose the probability P that a vector $\underline{x}$ will fall inside a region R in $\underline{x}$-space is given by

$$P = \int_R p(\underline{x}')\, d\underline{x}' \tag{4.6}$$

If N samples are drawn independently from $p(\underline{x})$, then the probability that K of them will fall within the region R is given by the binomial law

$$P(K) = \binom{N}{K} P^K (1 - P)^{N-K} \tag{4.7}$$

Since the mean fraction of the samples falling within this region is given by $E[K/N] = P$ and the variance about this mean is given by $E\left[(K/N - P)^2\right] = P(1-P)/N$, the distribution peaks sharply as $N \to \infty$. Thus, the mean fraction of the samples falling within the region R is a good estimate of the probability P so that

$$P \cong K/N. \tag{4.8}$$

Furthermore, if we assume that $p(\underline{x})$ is continuous and that the region R is small enough so that $p(\underline{x})$ does not vary appreciably within it, we have

$$P = \int_R p(\underline{x}')\, d\underline{x}' \cong p(\underline{x}) V \tag{4.9}$$

where V is the volume enclosed by the region R and $\underline{x}$ is an arbitrary point within R. Combining Eqs. 4.8 and 4.9 we obtain the following estimate for our probability density function $p(\underline{x})$,

$$\hat{p}(\underline{x}) = \frac{K}{NV}. \tag{4.10}$$

An appropriate nonparametric method for implementing our density estimation in Eq. 4.10 is the KNN technique. With the KNN density estimation, the approach is to select an appropriate K and determine the volume V containing the K samples centered on the point $\underline{x}$. Thus, the volume V is a function of the training data. Consequently, if the density of the training data is high near $\underline{x}$, the volume will be relatively small, leading to good resolution. On the other hand, if the density is low, the volume will grow until it obtains the required number of K, but it may stop growing sooner if it enters a region of higher density. Theoretically, to ensure $\hat{p}(\underline{x})$ is a good estimate of the probability that the point $\underline{x}$ will fall within the region R of volume V we desire K to approach infinity as N approaches infinity. However, to ensure that V shrinks to zero we must require K to approach infinity slower than N. Devroye, Györfi, and Lugosi [13] show that $\lim_{N \to \infty} K = \infty$ and $\lim_{N \to \infty} \frac{K}{N} = 0$ are necessary and sufficient conditions for $\hat{p}(\underline{x})$ to be a consistent estimate of $p(\underline{x})$.

Now suppose our training data set consists of N_j feature vectors from object class O_j and there are $\sum_{j=1}^{J} N_j = N$ points in total. As displayed in Fig. 4.9, we can draw a hypersphere of volume V with a center feature vector $\underline{D}_n$ and consisting of K other feature vectors irrespective of their object class. Suppose the hypersphere contains K_j feature vectors from object class O_j. From results of the probability density function estimation above in Eq. 4.10, we obtain our required likelihood function as a KNN density estimation

$$\hat{P}(\underline{D}_n \mid O_j) = \frac{K_j}{N_j V} \tag{4.11}$$

The underlying concepts for using the KNN density estimation in nonparametric discrimination originated from the works by Fix and Hodges [23, 24]. Their decision rule was to assign $\underline{D}_n$ to class j if

$$\hat{P}(\underline{D}_n \mid O_j) > \hat{P}(\underline{D}_n \mid O_i), \quad i \neq j \tag{4.12}$$

Fig. 4.9 K-Nearest-Neighbor density estimation.

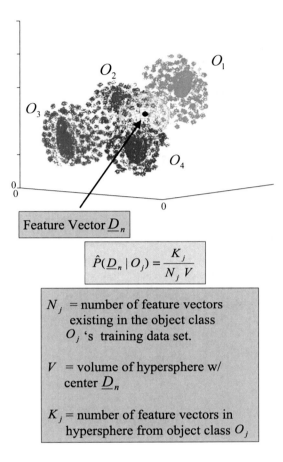

$$\hat{P}(\underline{D}_n | O_j) = \frac{K_j}{N_j V}$$

N_j = number of feature vectors existing in the object class O_j's training data set.

V = volume of hypersphere w/ center $\underline{D}_n$

K_j = number of feature vectors in hypersphere from object class O_j

for two classes $i = 1,2$. However, this maximum likelihood decision rule does not consider any prior knowledge about our object class O_j (i.e., $P(O_j)$). Therefore, our desired posterior probability for our Bayesian classifier in Eq. 4.3 is

$$P(O_j | \underline{D}_n) = \frac{\hat{P}(\underline{D}_n | O_j) P(O_j)}{\hat{P}(\underline{D}_n)}$$
$$= \frac{\frac{K_j}{N_j V} P(O_j)}{\hat{P}(\underline{D}_n)} \quad (4.13)$$

In our case, we will use a Bayes' decision rule as given in Eq. 4.4 that assigns the feature vector $\underline{D}_n$ to the object class with the maximum posterior probability.

At this point, two comments need to be made. First, as discussed above in the derivation of Bayes' theorem, the likelihood function is not necessarily a probability density function since the integral of a likelihood function may not equal to one. Thus, the KNN density function is not a true probability density since if we integrated Eq. 4.11 over the whole feature space, we would find that the integral is infinite (rather than unity). Second, in practice, the optimal value of K depends on the size of the available training data set and various approaches are used to determine the best value for K that results in the most favorable classifier performance. Consequently, the performance of the Bayesian classifier with a KNN density estimation depends on both the choice for K and the feature vector. We will discuss our choices for K in Sect. 4.4.4 below.

4.4.2 K-Nearest-Neighbor (KNN) Classifier

The traditional K-Nearest-Neighbor classifier (or rule) originated from the works of Cover and Hart [25]. They assumed that the proportion of each object class's samples in the training data set provides a good representation for the prior probability $P(O_j)$ of that object class being present in the environment for subsequent classifications so that

$$P(O_j) = \frac{N_j}{N}. \tag{4.14}$$

In this case, the unconditional density (evidence) in Eq. 4.2 becomes

$$\begin{aligned}
\hat{P}(\underline{D}_n) &= \hat{P}(\underline{D}_n, O_1) + \hat{P}(\underline{D}_n, O_2) + \ldots + \hat{P}(\underline{D}_n, O_J) \\
&= \sum_{j=1}^{J} \hat{P}(\underline{D}_n \mid O_j) P(O_j) \\
&= \sum_{j=1}^{J} \frac{K_j}{N_j V} \left(\frac{N_j}{N} \right) \\
&= \sum_{j=1}^{J} \frac{K_j}{NV} \\
&= \frac{\sum_{j=1}^{J} K_j}{NV} \\
&= \frac{K}{NV}.
\end{aligned} \tag{4.15}$$

For the posterior probability we have

$$P(O_j \mid \underline{D}_n) = \frac{\hat{P}(\underline{D}_n \mid O_j) P(O_j)}{\hat{P}(\underline{D}_n)}$$

$$= \frac{\dfrac{K_j}{N_j V}\left(\dfrac{N_j}{N}\right)}{\dfrac{K}{NV}} \tag{4.16}$$

$$= \frac{K_j}{NV}\left(\frac{NV}{K}\right)$$

$$= \frac{K_j}{K}.$$

This form leads to what has traditionally been known as the K-Nearest-Neighbor classifier (or rule). Generalizing to M classes, we assign $\underline{D}_n$ to class j if

$$K_j > K_i, \quad j \neq i, \quad i = 1, 2, \ldots, M. \tag{4.17}$$

Consequently, the design rule is to assign $\underline{D}_n$ to the class that receives the majority vote amongst the K nearest neighbors. The case where $K = 1$ is simply called the Nearest Neighbor Rule.

4.4.3 Parzen Classifier

The Parzen classifier estimates the object class densities by a Parzen density estimation [26]. Both the KNN and Parzen density estimations evolve from Eq. 4.10. With the KNN density estimation presented in Sect. 4.4.1, the volume V of the hypersphere with a center feature vector $\underline{D}_n$ is determined by the specified number of nearest neighbors K that depends on the size N of the training data set. However, the Parzen density estimation reverses the roles. In the Parzen density estimation, the value of K is determined by a specified volume V that depends on the size N of the training data. Similar to the Bayesian classifier with the KNN density estimation, the Parzen classifier will estimate the densities for each object class and assign an unknown feature vector to the object class with the maximum posterior probability.

Beginning with Eq. 4.10, suppose our training data set consists of N_j feature vectors from object class O_j and there are $\sum_{j=1}^{J} N_j = N$ vectors in total. Draw a d-dimensional hypercube with edges of length h and a center feature vector $\underline{D}_n$ around K other feature vectors irrespective of their object class. The hypercube contains K_j feature vectors from object class O_j. The volume of this hypercube is given by

$$V = h^d \tag{4.18}$$

We can derive an analytical expression for K_j by defining the kernel function (or Parzen window function):

$$H(\underline{u}) = \begin{cases} 1 & |u_p| \leq 1/2 \quad p = 1,...,d \\ 0 & otherwise. \end{cases} \tag{4.19}$$

Thus, the Parzen density estimation is known as a kernel-based method for density estimations. The function $H(\underline{u})$ defines a unit hypercube centered at the origin. Consequently, for all feature vectors $\underline{D}_{qj}$ from the training data set of object class O_j, the value of $H((\underline{D}_n - \underline{D}_{qj})/h)$ is unity if the point $\underline{D}_{qj}$ falls within the hypercube and is zero otherwise. The total number of feature vectors from object class O_j in the hypercube is given by

$$K_j = \sum_{q=1}^{N_j} H\left(\frac{\underline{D}_n - \underline{D}_{qj}}{h}\right) \tag{4.20}$$

By substituting Eqs. 4.18 and 4.20 into Eq. 4.10, we obtain our Parzen density estimation:

$$\hat{P}(\underline{D}_n | O_j) = \frac{1}{N_j} \sum_{q=1}^{N_j} \frac{1}{h^d} H\left(\frac{\underline{D}_n - \underline{D}_{qj}}{h}\right) \tag{4.21}$$

Therefore, our posterior probability for our Parzen classifier is given by

$$P(O_j | \underline{D}_n) = \frac{\hat{P}(\underline{D}_n | O_j) P(O_j)}{\hat{P}(\underline{D}_n)}$$

$$= \left[\frac{1}{N_j} \sum_{q=1}^{N_j} \frac{1}{h^d} H\left(\frac{\underline{D}_n - \underline{D}_{qj}}{h}\right)\right] \frac{P(O_j)}{\hat{P}(\underline{D}_n)} \tag{4.22}$$

The performance of the Parzen classifier depends on both the choice for h and the feature vector. We will discuss our method for choosing h in Sect. 4.4.4 below.

4.4.4 General Remarks

In this section, we will make some general remarks that are common to all our classifiers. First we will comment on the choices for K used in the KNN density estimation and h used by the Parzen density estimation. Next we will comment on the use of prior probabilities by the classifiers. We will conclude the section with a brief discussion on how to deal with ties between two posterior probabilities with different class assignments.

4.4.4.1 Choices for Parameters K and h

Both K and h act as smoothing parameters for the KNN and Parzen density estimations, respectively, where an appropriate choice will result in a good approximation to the true density function for the training data. However, for our non-parametric application where the density function is not known, we must choose parameter values that minimize the misclassification error. Two approaches for selecting the values of the parameters are by either presenting the parameters as a function of the training data or using cross-validation.

Since the KNN density estimation is one of the most popular methods used in pattern classification, there exists a considerable amount of research in the literature to develop a scheme for choosing the value for the parameter K that will minimize the misclassification rate [27, 28, 29, 30]. For our Bayesian classifier, we will choose the following functional form of K in terms of the training data size that was presented by Loftsgaarden and Quesenberry [27] and endorsed by Duda, Hart, and Stork [1]:

$$K(N_j) = \sqrt{N_j} \qquad (4.23)$$

where N_j is the number of labeled observations in the training data set for object class O_j. A functional form for the parameter h in the Parzen density estimation that is recommended by Duda, Hart, and Stork [1] is obtained by letting $V(N_j) = 1/\sqrt{N_j}$ in Eq. 4.18. We will choose the parameters for the KNN and Parzen classifiers using the cross-validation method discussed below.

Cross-validation is an error estimation method used to assist in designing a classification model with a minimum misclassification error. The most favorable pattern classification model is the one consisting of the feature vector and parameters in the classifier that results in minimal classification errors. The classifier is first designed using a training data set, and then its classification performance is assessed using a test (or validation) set. Hence, the test set is used to tune the values of the parameters in the classifier. The percentage of misclassified test samples is used as an estimate of the error (or misclassification) rate. Thus, cross-validation is used to compute the error rate for different parameter values (i.e., k or h) for a classifier and a given feature vector. The parameter value that results in the lowest estimate of the error rate is chosen for the given classifier. We will use the cross-validation method to select the parameter values for the KNN and Parzen classifiers. A more detailed discussion on the cross-validation method will be provided in Sect. 4.5.

4.4.4.2 Prior Knowledge

The quantity $P(O_j)$ in Bayes' formula (Eq. 4.3) is called the prior probability since it represents our state of knowledge regarding the object class being present before any signal data is collected using the sensor. Our Bayesian and Parzen classifiers possess the capability to input prior knowledge regarding each object class's existence in the robot's local area of operation. However, as we mentioned previously, the KNN classifier assumes that the proportion of each object class's samples in the training data set provides a good representation for the prior probability.

The KNN classifier's prior probability may be appropriate when dealing with training data sets that form compact hyperspherical clusters. However, the KNN classifier's prior probability may not be appropriate with our hyperconoidal clusters where multimodal distributions normally occur within each object class due to the dynamical outdoor environment's effect on the training data. For instance, multimodal distributions could occur within an object class's training data set since the features generated from the object's thermal images display variations in the values due to the diurnal cycle of solar energy. Furthermore, since we are seeking to classify objects that could exist in a robot's local area of operation, we may want to incorporate a prior based on our knowledge of an object existing in the environment under inspection. For example, if we feel that all the object classes could exist in our robot's local area of operation or have no reason to believe that one object class is more likely to be identified over another, then the "principle of indifference" prevails and we assign equal priors for all the object classes.

During our analysis in the present chapter and Chap. 5, we will assume equal prior probabilities for all our object classes when using the Bayesian and Parzen classifiers. We will also use the popular KNN classifier as the comparative benchmark regardless of its potential shortcoming with the prior probability. In Chap. 6, we will discuss future research to assign a prior probability to an object class using knowledge gained from satellite imagery.

4.4.4.3 Ties

There are various approaches to deal with ties between two posterior probabilities with different class assignments. For the KNN classifier, Devroye, Györfi, and Lugosi, [13] recommend choosing K to be odd to avoid voting ties. Webb [7] provides several ways to break ties. One way is to break ties arbitrarily. Another possible tiebreaker technique is to assign $\underline{D}_n$ to the object class, out of the classes with equal posterior probabilities, that has the nearest mean vector to $\underline{D}_n$ (where the mean vector is computed over each object class's training data within the cell of volume V). An alternative method is to assign $\underline{D}_n$ to the most compact object class out of the classes with equal posterior probability values. Since our autonomous robot may have to decide whether to go through the hedge or around the brick wall, posterior probabilities for the hedge and brick wall that are close in value could result in an autonomous robot with damaged sensors if the brick wall was misclassified as a hedge. Our point of view is that two posterior probabilities with different recommendations for class assignments but a small absolute difference in their posterior values may present too much risk for a misclassification. Consequently, in Chap. 5, we will introduce our approach that will prevent ties and high-risk decisions by requiring the two highest posterior values with different recommendations for class assignment to have an absolute difference that exceeds a specified threshold value. If the rule for the threshold is not satisfied, the classification is rejected and the robot must capture another thermal image for class assignment.

4.5 Model Performance and Feature Selection

In this section we will discuss and implement methods to select the most favorable feature vectors that result in minimum classification errors for the Bayesian, KNN, and Parzen classifiers presented in Sect. 4.4. Since the performance of a classifier is a function of the feature vector as well as the value of the its parameters (i.e., K or h), the evaluation of classifiers and selection of feature sets are done simultaneously using various error estimation methods. Our selection process will involve an exhaustive search using two high performance computing systems to analyze the classification performance of over 290,000 feature combinations spanning up to 18 dimensions. A login node was used on the DoD High Performance Computing Modernization Program system at the Army Research Laboratory Major Shared Resource Center that included 8 GB of memory at a processor frequency of 3.6 GHz. Four nodes were used on a computing system located at the College of William & Mary with two nodes each consisting of 8 GB of memory and two other nodes each consisting of 16 GB of memory, each node operating at a processor frequency of 1.28 GHz. The end-state objective is to present sets of features that will maximize the performance of classifiers in assigning the correct object class to unknown feature vectors generated from the thermal imagery of non-heat generating objects in an outdoor environment.

The discussions in this section are outlined as follows. In Sect. 4.5.1 we will discuss our exhaustive search feature selection method. Section 4.5.2 will present our performance criteria used to assess each classification model (classifier plus feature vector). Section 4.5.3 will discuss our error estimation methods used on the training and test data sets. In Sect. 4.5.4 we will provide a summary of our process for evaluating the various classification models and selecting the most favorable feature vectors for our extended and compact objects. We will select the most favorable feature vectors for the extended objects in Sect. 4.5.5 and compact objects in Sect. 4.5.6.

4.5.1 Feature Selection Method

This section is concerned with the method used to identify the most favorable features for classifying an unknown non-heat generating object with minimal error. We will discuss the two primary approaches used to identify these features – *feature selection and feature extraction*. Our discussion will include how and why we use a feature selection method to identify our most favorable sets of features and feature extraction method in a "nontraditional way" to analyze the hyperconoidal clusters and design our novel classification model in Chap. 5. A review of these two methods is found in [2, 3, 7]. The goal of both methods is to minimize both the number of dimensions of the features and misclassifications. Not only does a large dimensional feature vector, relative to the available training data, increase the computation time for the robot's decision-making process but, more importantly, it will have a negative effect on the performance of the classification model. This behavior brings up the concepts of the *curse of dimensionality* and *peaking phenomena* that we will discuss first.

According to the *curse of dimensionality* [4], as the number of dimensions increases for a feature vector, the size of the training data set must increase exponentially as a function of the feature dimension to obtain an increase in classification performance. However, in practice, we have a limited quantity of data. Thus, as the number of dimensions of a feature vector increases, the data becomes sparse, in which case the classification performance begins to decline. This behavior is known as the *peaking phenomenon* [2]. Consequently, a rule of thumb that we will adopt to favor peak performance of our classification model is to have no more than $n/10$ features for an object class with n training patterns [2].

4.5.1.1 Feature Extraction

Feature extraction methods create new features based on transformations of the original feature set. Thus, the new feature set may not have a clear physical meaning or retain the physical interpretation found in the original features generated from an object's thermal image. Consequently, we will not use any feature extraction methods in the "traditional way" for our application. Some of the popular feature extraction methods include principal component analysis, Karhunen-Loève transformation, independent component analysis, factor analysis, discriminant analysis (also known as Fisher linear discriminant analysis), and multi-dimensional scaling [2, 3, 7, 31, 32]. We will only discuss principal component analysis since we will apply this method in a "nontraditional way" when we analyze the hyperconoidal clusters and introduce the design of our novel classification model in Chap. 5.

4.5 Model Performance and Feature Selection

Principal component analysis (PCA) is traditionally applied to the entire feature space in unsupervised classification. The objective is to transform the original features to a lower dimensional space while retaining as much information about the original features as possible. The idea behind this method is that the information in the patterns of an n-dimensional feature space can be represented by a transformation involving the projection of the patterns, irrespective of any object class information, onto a subset of n orthonormal vectors with directions corresponding to high variance in the patterns. PCA assumes that information about the original features is available in the variance of the features. Hence, a direction of higher variance in the patterns corresponds to more information about the features. Any vector in a direction of low variance can be excluded from the transformation since it implies a direction with a low amount of information about the features. Thus, the projection of the original patterns onto each of the selected vectors in the directions of the highest variance will yield new patterns in a lower dimensional space. In some cases the resulting transformation could yield an acceptable separation of the original clusters in the feature space. For instance, in Fig. 4.10, the projection of the 2-dimensional patterns onto the vector $\underline{e}_1$ in the di-

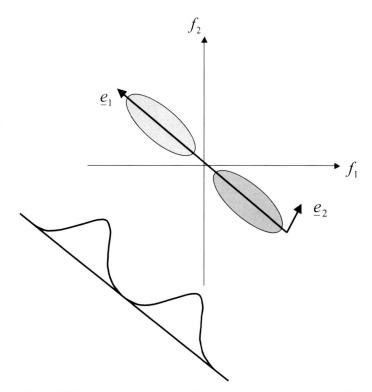

Fig. 4.10 Principal component analysis used to project patterns onto eigenvector in direction of maximum variance of the patterns.

rection of the maximum variance of the patterns and excluding the vector $\underline{e}_2$ from the transformation would reduce the patterns to a 1-dimensional feature space while providing an acceptable separation of the two clusters as indicated by their given distributions.

We now go into more detail on the derivation of the transformation used in PCA. Let F be an $n \times m$ training data matrix where each column forms an n-dimensional feature vector $f = \langle f_1, f_2, ..., f_n \rangle$ for one object. First center the data by subtracting the sample mean $\overline{f_i}$ from the feature value f_i, where $i = 1,...,n$, across each row in F. This produces a matrix $\widetilde{F}$ where $\widetilde{f_i} = f_i - \overline{f_i}$ so that each row has a mean of zero. Compute the covariance matrix C of the centered training data matrix $\widetilde{F}$ so that

$$C\{\widetilde{F}\} = \begin{bmatrix} \sigma^2\{\widetilde{f_1}\} & \sigma\{\widetilde{f_1},\widetilde{f_2}\} & \cdots & \sigma\{\widetilde{f_1},\widetilde{f_n}\} \\ \sigma\{\widetilde{f_2},\widetilde{f_1}\} & \sigma^2\{\widetilde{f_2}\} & \cdots & \sigma\{\widetilde{f_2},\widetilde{f_n}\} \\ \vdots & \vdots & \ddots & \vdots \\ \sigma\{\widetilde{f_n},\widetilde{f_1}\} & \sigma\{\widetilde{f_n},\widetilde{f_2}\} & \cdots & \sigma^2\{\widetilde{f_n}\} \end{bmatrix}. \tag{4.24}$$

The covariance matrix C is an $n \times n$ matrix with the variances of the individual features of $\widetilde{F}$ along the main diagonal and the off-diagonal elements consist of the covariances of each pair of features. Since $\sigma\{\widetilde{f_i},\widetilde{f_j}\} = \sigma\{\widetilde{f_j},\widetilde{f_i}\}$ for all $i \neq j$, $C\{\widetilde{F}\}$ is a symmetric matrix. From linear algebra, the matrix $C\{\widetilde{F}\}$ is symmetric if and only if it has an orthonormal set of n eigenvectors. Next, calculate the eigenvalues λ_i and corresponding unit-length eigenvectors $\underline{e}_i$ of the covariance matrix in Eq. 4.24 in following algebraic eigenvalue problem

$$C\underline{e}_i = \lambda_i \underline{e}_i. \tag{4.25}$$

Thus, a large eigenvalue λ_i equates to large covariance values (positive or negative) for pairs of features in C. The unit-length eigenvector $\underline{e}_i$ corresponding to this large eigenvalue provides a direction of high variation in the patterns. The ordering of the eigenvectors is such that the corresponding eigenvalues λ_i satisfy $\lambda_1 \geq \lambda_2 \geq ... \geq \lambda_n$. The largest eigenvalues λ_1 is associated with the eigenvector $\underline{e}_1$ that determines the direction with the maximal variance and best fits the patterns in a least squared sense.

Each eigenvector in the orthonormal set corresponds to a *principal axis* of the patterns in the feature space. The PCA transformation projects each pattern

$\underline{\tilde{f}} = \langle \tilde{f}_1, \tilde{f}_2, \ldots, \tilde{f}_n \rangle$ onto a given column eigenvector $\underline{e}_j$ to obtain a new feature given by the linear combination

$$y_j = \sum_{i=1}^{n} e_{ij} \tilde{f}_i \qquad (4.26)$$
$$= \underline{e}_j^T \underline{\tilde{f}}$$

where j is the index of the chosen eigenvector in a direction with a high variance of the patterns. Each of these new features y_j given by Eq. 4.26 is called a *principal component*. The principal component y_1 corresponding to the eigenvector $\underline{e}_1$ in the direction of the maximal variance of the patterns is called the first principal component.

As mentioned earlier, we can choose a subset of the eigenvectors with directions corresponding to the highest variances of the patterns and exclude those eigenvectors with directions of low variance to obtain a new lower dimensional pattern with minimal loss of information about the original features. Let E denote a matrix with each column being one of the selected eigenvectors $\underline{e}_j$. The PCA transformation

$$Y = E^T \tilde{F} \qquad (4.27)$$

yields new patterns Y, consisting of principal components, in a reduced dimensional feature space. Each column of Y is a new lower dimensional feature vector corresponding to the same column with the original feature vector in $\tilde{F}$. Consequently, if we included all n eigenvectors in E, we would lose no information, and Y would contain the original data rotated in the feature space with the eigenvectors as the axes. We will not use PCA as a feature extraction method; however, we will use its ability to fit an eigenvector through an object class's hyperconoidal cluster in a least squares sense. Using PCA in this "nontraditional way" will allow us to analyze the characteristics of the hyperconoidal clusters for each object class and assist in designing our novel classification model in Chap. 5.

4.5.1.2 Feature Selection

Contrary to the feature extraction methods, feature selection methods result in features sets that retain their original physical meaning. The process in feature selection methods is to select the subset of size d from the available input feature set of size p that leads to the most favorable performance for a specific classifier based on a given criterion $J(\cdot)$. We will discuss our choice for a performance criterion in Sect. 4.5.2. Since the most favorable subset of features is dependent on the type of classifier chosen, the selected features are "wrapped around" the given classifier. Consequently, feature selection methods are often referred to as "wrapper methods" [3].

The most popular feature selection methods include exhaustive search, branch-and-bound search, best individual features, sequential forward selection, sequential backward selection, plus l-take away r selection, sequential forward floating search, and sequential backward floating search [2]. The exhaustive search is the only thorough approach to identifying the most favorable feature vector since it involves examining all $\binom{p}{d}$ possible subsets and selecting the subset that leads to the best performance for a specific classifier based on the criterion $J(\cdot)$. As noted in [2], no nonexhaustive feature selection method can be guaranteed to produce the "optimal" subset. The exhaustive search is normally avoided since it is computationally expensive. However, with the increasing capabilities of high performance computing systems, what used to take say 20 days to evaluate 32,000 combinations of feature subsets, currently takes 4 days to complete. Therefore, our approach is to use the exhaustive search feature selection method on the high performance computing systems that we discussed earlier. As mentioned previously, to ensure peak performance, the size d of this most favorable feature vector must also satisfy the rule of thumb to have no more than $n/10$ features for an object class with n training patterns.

4.5.2 Performance Criterion

The most favorable classification model (feature vector plus classifier along with parameter values) is determined by comparing performance criterion values for all possible combinations of features and classifiers by an exhaustive search. Choices for the performance criterion functions $J(\cdot)$ normally include the estimated misclassification (or error) rate P_e, estimated correct classification (or accuracy) rate $(1-P_e)$, or some distance measure as the performance criterion $J(\cdot)$. For our application, we seek to determine the classification model that minimizes the estimated error rate criterion given by

$$P_e = \frac{n_e}{n} \qquad (4.28)$$

where n_e is the number of misclassified feature vectors out of n labeled test set samples for a given object class. The criterion based on a distance measure normally consist of a ratio of the distances between object class clusters (interclass) and feature vectors within each object class (intraclass) [7]. The interclass/intraclass distance criterion should show a strong linear relationship with the estimated error rates such that the interclass/intraclass value increases as the estimated error rate decreases in value. However, we investigated the use of the interclass/intraclass distance criterion in our application and found a weak relationship between the estimated error rates and the interclass/intraclass distances. The best coefficient of determination of $r^2 = 57.4\%$ was achieved with a Bayesian classifier and 2-dimensional feature vectors. The coefficients of determination decreased in value as the feature vectors increased in dimensions. Consequently, this type of distance criterion is best for applications involving hyperspherical or hyperellipsoidal clusters with no common origin amongst the object classes, as we see in our application involving hyperconoidal clusters.

In Sect. 4.5.3, we will discuss our chosen error estimation methods that involve the use of training data to design a classifier and test (or validation) data to assess the performance of the classification model. For a given classification model, these methods will assign an object class label to each feature vector from a test data set consisting of known (or actual) labels from multiple object classes. The resulting class assignments of the test data set by the given classification model will be presented in a *confusion matrix* (or misclassification matrix). Table 4.1 provides an example of a confusion matrix involving the extended objects where the labels for the actual object classes are displayed along the columns and the labels for the assigned object classes are given along the rows. Each element of the matrix, given by the ith row and jth column, provides the number of feature vectors from the actual object class ω_j that were assigned as object class ω_i by the given classification model. For example, out of the 23 actual brick wall feature vectors in the test data set, the classification model correctly assigned 15 feature vectors as brick wall and misclassified 6 feature vectors as hedges and 2 feature vectors as wood walls. By applying Eq. 4.28, the error rate for the brick wall is approximately 34.78%. The error rates for each object class are displayed below the confusion matrix. When comparing the performance of all the classification models using the exhaustive search feature selection method we will use the average of the error rates for each object class in the test data set due to the large number of models being evaluated. The average error rate for our example in Table 4.1 is approximately 33.70%. Once we identify the most favorable feature vectors, we will use the more detailed error rates for each object class in the confusion matrix during our analysis and design of our most favorable classification model in Chap. 5.

Table 4.1 Confusion matrix example that assesses a classification model's performance on test data set consisting of extended objects.

			Actual Object Class			
		Object Labels	Brick Wall	Hedges	Picket Fence	Wood Wall
			1	2	3	4
Assigned Object Class	Brick Wall	1	15	3	0	5
	Hedges	2	6	20	0	8
	Picket Fence	3	0	0	20	4
	Wood Wall	4	2	0	3	6
	Total in Object Class		23	23	23	23
	Total Errors		8	3	3	17
	Error Rate (%)		34.7826	13.0435	13.0435	73.9130
	Total Errors		31			
	Avg Error Rate (%)		33.6957			

4.5.3 Error Estimation Method

In this section we will discuss our choice of error estimation methods that involve the use of training data to design a classifier and test (or validation) data to assess the performance of the classification model. The training, test, and blind data sets used in our application were discussed in Sect. 2.3. The objective of the error estimation methods is to manage the training and test data sets that are used by a given classification model to ensure an appropriate estimation of the error rate. The error estimation methods that are commonly used in pattern classification include the resubstitution method, holdout method, leave-one-out method, rotation method, and bootstrap method [1, 2, 5, 6, 7]. The holdout, leave-one-out, and rotation methods are different versions of the cross-validation algorithm [2]. For each classification model (i.e., classifier plus feature vector), an average error rate is computed on the test set data using a given error estimation method. In our application, we will estimate the average error rates using the resubstitution method, holdout method, and leave-one-out method.

In the resubstitution method, all the available data used for the training data set is also used as test data. The resubstitution method will only be applied to the Bayesian classifier. Thus, for this method, our training data collected from 15 March to 22 June 2007 will be used in Sects. 4.5.5 and 4.5.6 to design the Bayesian classifier with a given feature vector and then resubstituted as test data to validate the design.

In the holdout method, a portion of the data is used for training and another portion is used for testing. Thus, the training and test data sets are disjoint. In this case, the training set is the data collected from 15 March to 22 June 2007. We will use the test set collected from 25 June to 3 July 2007 in Sects. 4.5.5 and 4.5.6 to assess the performance of the Bayesian, KNN, and Parzen classifiers. We will use our blind data set that was collected from 6 July to 5 November as our validation set when we analyze our most favorable feature vectors and designing our novel classification model in Chap. 5.

The leave-one-out method uses the training set of size N to design the classifier using $(N-1)$ samples as the training data and assess the classifier on the one remaining feature vector as the test sample. This process is repeated N times with different training sets of size $(N-1)$ to compute an average estimated error rate. We will apply the leave-one-out method to compute the average error rates involving the KNN and Parzen classifiers in Sects. 4.5.5 and 4.5.6. As discussed in Sect. 4.4.4.1, cross-validation is also used to identify the most favorable parameter value for a given classifier. The parameter value that results in the lowest estimate of the average error rate is chosen for the given classifier. Therefore, we will also use the leave-one-out method to select the parameter value for K in the KNN classifier and h in the Parzen classifier. The leave-one-out method will use the training data collected from 15 March to 22 June 2007. The leave-one-out method is also called the jackknife method by John W. Tukey since it is handy and useful in many ways [1].

4.5.4 Checkpoint Summary

In Sects. 4.5.5 and 4.5.6 below, we will evaluate the performance of various classification models and identify the most favorable feature vectors for our extended and compact objects, respectively. In this section we will summarize our process used in the following two sections that is based on the concepts we discussed in Sects. 4.4 through 4.5.3. The goal of Sects. 4.5.5 and 4.5.6 is to select a set of feature vectors that result in the lowest error rates when teamed up with either the Bayesian, KNN, or Parzen classifier. We will compute the error rates for each classifier combined with every combination of features across all possible dimensions (i.e., exhaustive search feature selection method). For each classification model (i.e., classifier plus feature vector), an average error rate is computed on test set data using the resubstitution, holdout, and leave-one-out error estimation methods. The resulting average error rates are compared to determine the feature vectors that present the lowest error rates. These feature vectors will be considered as our most favorable feature vectors and used for further analysis and designing our novel classification model in Chap. 5.

4.5.5 Extended Object Performance and Feature Selection

The 18 thermal features remaining from our preliminary feature analysis in Sect. 4.3 are displayed in Table 4.2 along with numerical labels that are provided for convenience as we analyze the different feature vectors used in the classification models during the exhaustive search feature selection method. The equations for each feature were discussed and derived in Chap. 3.

Table 4.3 provides the number of combinations of extended object features for each feature vector dimension used in the exhaustive search method. We will compute the average error rates for the classification models across all 18 dimensions to ensure an exhaustive search. However, we will also adhere to rule of thumb given in Sect. 4.5.1 that requires the size d of the most favorable feature vector to have no more than $n/10$ features for an object class with n training patterns to ensure peak performance. Thus, given the number of training patterns for each extended object class in Table 2.1, the maximum acceptable size for our most favorable feature vector is 11 features.

The average error rates for each classifier combined with every combination of features across all possible dimensions (i.e., exhaustive search feature selection method) were computed using the error estimation methods on a high performance computing system discussed in Sect. 4.5. Figure 4.11 presents dotplots that give the general trend of the average error rates for each classifier and error estimation method observed in each dimension. The dotplots show that holdout error estimation methods have a tendency to display a higher variance in the average error rates compared to the leave-one-out and resubstitution methods. This oc-

Table 4.2 Extended object thermal features and labels used in the exhaustive search feature selection method.

FEATURE	LABEL
Meteorological Features	
Ambient Temp. $°F$ (Ta)	1
Amb. Temp. Rate of Change (T1)	2
Micro Features	
Object Surface Radiance (Lo)	3
Reference Emitter Radiance (Lr)	4
Background Irradiance (Lb)	5
Lo/Lr (Lor)	6
Lo/Lob (Lob)	7
Emissivity (Eo)	8
Macro Features	
First-order Statistics	
Object Scene Radiance (Mo1)	9
Mo1/Lr (Mor1)	10
Mo1/Lb (Mob1)	11
Smoothness (So1)	12
Entropy1 (En1)	13
Second-order Statistics	
Contrast2 (Co2)	14
Correlation (Cr2)	15
Energy (Er2)	16
Homogeneity (Ho2)	17
Entropy2 (En2)	18

curs since the training and test data come from the same set for the leave-one-out and resubstitution methods.

The average error rates were sorted in increasing order by classifier and error estimation method within each dimension. Tables 4.4 (a–e) compare the lowest average error rates (%) of each classifier with the respective error estimation method across each feature vector dimension. The average error rates in Tables 4.4 (a–e) clearly illustrate the behavior known as the peaking phenomenon. Thus, as the number of dimension of a feature vector increases, the error rates of each classifier decrease to a specific peak (or in some cases a short plateau) and then the classification performance begins to decline. For instance, the Bayesian classifier with the resubstitution error estimation method reaches its peak performance at an estimated average error rate of 16.70% with a 7-dimensional feature vector. We also see that no classifier reaches a peak performance with a feature vector consisting of only features from a single feature category – meteorological, micro, or macro.

4.5 Model Performance and Feature Selection

Table 4.3 Total number of extended object thermal feature combinations for feature vectors from 1 to 18 dimensions. The first 11 dimensions (highlighted in yellow) satisfy the rule of thumb to ensure peak performance of the classification models.

DIMENSIONS OF FEATURE VECTOR	NUMBER OF FEATURE COMBINATIONS
1	18
2	153
3	816
4	3060
5	8568
6	18564
7	31824
8	43758
9	48620
10	43758
11	31824
12	18564
13	8568
14	3060
15	816
16	153
17	18
18	1
TOTAL	**262143**

The next step is to compare classification models (along with their respective error estimation methods) in Tables 4.4 (a–e) to identify a most favorable set of feature vectors. The size of the most favorable set of feature vectors is limited to 11 dimensions to support the rule of thumb for peak performance. We choose pairs of classification models for comparison based on their similarities in the error rate trends found in Fig. 4.11. Thus, for each dimension, we compare the error rates in Table 4.4 for the

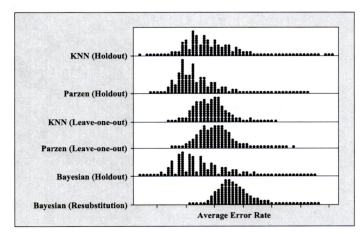

Fig. 4.11 General trend for extended objects of dotplots with average error rates for each classifier and error estimation method observed in each dimension.

Table 4.4 Extended object comparison of the lowest average error rates (%) of each classifier with the respective error estimation method across each feature vector dimension.

(a)

FEATURE VECTOR					CLASSIFIER	ERROR ESTIMATION METHOD	AVERAGE ERROR RATE (%)	K VALUE	h VALUE
3					Bayesian	Resubstitution	44.3569	*	
6					Bayesian	Holdout	30.0347	*	
6					KNN	Holdout	30.0347	54	
3					KNN	Leave-one-out	46.2261	5	
6					Parzen	Holdout	31.5972		0.0199
3					Parzen	Leave-one-out	43.9564		0.0123
4	13				Bayesian	Resubstitution	30.0623	*	
6	16				Bayesian	Holdout	13.8889	*	
3	7				KNN	Holdout	19.6181	19	
1	12				KNN	Leave-one-out	30.3894	1	
5	13				Parzen	Holdout	17.8819		0.0433
10	13				Parzen	Leave-one-out	32.4522		0.0247
5	6	18			Bayesian	Resubstitution	23.9697	*	
5	6	16			Bayesian	Holdout	11.9792	*	
3	6	16			KNN	Holdout	9.3750	11	
1	2	12			KNN	Leave-one-out	10.3427	1	
4	6	16			Parzen	Holdout	9.2014		0.0276
6	11	13			Parzen	Leave-one-out	23.0441		0.0361
3	5	6	13		Bayesian	Resubstitution	22.0828	*	
5	6	8	16		Bayesian	Holdout	7.2917	*	
3	5	6	8		KNN	Holdout	8.8542	16	
1	2	4	14		KNN	Leave-one-out	12.0004	1	
3	6	10	13		Parzen	Holdout	7.8125		0.0381
1	2	10	13		Parzen	Leave-one-out	16.9470		0.0381

(b)

FEATURE VECTOR							CLASSIFIER	ERROR ESTIMATION METHOD	AVERAGE ERROR RATE (%)	K VALUE	h VALUE
2	3	5	6	18			Bayesian	Resubstitution	19.0587	*	
3	5	6	9	16			Bayesian	Holdout	5.7292	*	
3	4	7	8	13			KNN	Holdout	6.0764	1	
1	2	3	6	12			KNN	Leave-one-out	10.6030	1	
1	3	6	7	18			Parzen	Holdout	6.2500		0.0394
1	2	10	11	13			Parzen	Leave-one-out	13.4335		0.0392
2	3	6	11	15	18		Bayesian	Resubstitution	18.7939	*	
3	5	8	9	10	17		Bayesian	Holdout	4.3403	*	
1	3	6	11	12	13		KNN	Holdout	6.0764	1	
1	2	3	6	12	14		KNN	Leave-one-out	10.5986	1	
1	3	5	7	10	16		Parzen	Holdout	5.9028		0.0536
1	2	3	4	12	14		Parzen	Leave-one-out	12.9662		0.0215
2	3	5	6	9	13	18	Bayesian	Resubstitution	16.7045	*	
3	5	6	8	9	13	16	Bayesian	Holdout	2.9514	*	
1	3	4	6	8	11	18	KNN	Holdout	4.6875	1	
1	2	4	5	6	10	18	KNN	Leave-one-out	11.5421	1	
1	3	5	6	7	11	16	Parzen	Holdout	4.3403		0.0534
1	2	4	5	6	10	18	Parzen	Leave-one-out	11.7713		0.0347
2	3	5	6	9	15	16	18	Bayesian	Resubstitution	16.9737	*	
3	5	6	7	8	9	13	16	Bayesian	Holdout	2.9514	*	
1	3	4	6	11	12	14	18	KNN	Holdout	4.5139	1	
1	2	3	4	5	6	10	18	KNN	Leave-one-out	11.0703	1	
1	3	5	6	7	12	14	16	Parzen	Holdout	5.9028		0.0565
1	2	4	5	6	10	14	17	Parzen	Leave-one-out	12.9617		0.0528

4.5 Model Performance and Feature Selection   127

(c)

FEATURE VECTOR														CLASSIFIER	ERROR ESTIMATION METHOD	AVERAGE ERROR RATE (%)	K VALUE	h VALUE
2	3	5	6	7	9	13	15	18						Bayesian	Resubstitution	16.9426	*	
2	3	5	6	7	8	9	11	12						Bayesian	Holdout	2.9514	*	
1	3	4	6	8	11	12	14	18						KNN	Holdout	3.1250	1	
1	2	3	4	5	6	10	14	18						KNN	Leave-one-out	11.3040	1	
1	3	5	6	10	12	13	14	17						Parzen	Holdout	6.2500		0.0536
1	2	3	4	5	6	8	10	13						Parzen	Leave-one-out	11.5287		0.0413
2	3	5	6	9	11	13	15	17	18					Bayesian	Resubstitution	17.6747	*	
2	3	5	6	8	9	13	16	17	18					Bayesian	Holdout	4.5139	*	
1	3	4	5	6	8	12	14	16	18					KNN	Holdout	6.0764	1	
1	2	3	4	5	6	10	12	14	18					KNN	Leave-one-out	11.5376	1	
1	3	4	5	6	10	12	13	14	17					Parzen	Holdout	6.2500		0.0509
1	2	3	4	6	10	11	12	13	14					Parzen	Leave-one-out	13.1820		0.0492
2	3	5	6	7	9	10	13	15	16	17				Bayesian	Resubstitution	17.1495	*	
1	2	3	5	6	8	9	10	13	16	17				Bayesian	Holdout	4.6875	*	
1	3	6	7	8	10	11	12	13	14	18				KNN	Holdout	6.0764	1	
1	2	3	4	5	6	8	10	12	14	18				KNN	Leave-one-out	12.0004	1	
1	3	4	5	6	8	10	12	13	14	17				Parzen	Holdout	6.2500		0.0590
1	2	4	5	6	7	10	11	12	14	16				Parzen	Leave-one-out	12.2786		0.0639
2	3	5	6	7	9	10	13	15	16	17	18			Bayesian	Resubstitution	18.1019	*	
1	2	3	6	7	8	9	11	13	15	16	17			Bayesian	Holdout	6.0764	*	
1	3	5	6	7	8	10	11	12	13	14	18			KNN	Holdout	7.6389	1	
1	2	3	4	5	6	8	10	11	12	13	14			KNN	Leave-one-out	12.2474	1	
1	3	4	5	6	7	8	10	12	14	15	16			Parzen	Holdout	9.0278		0.0598
1	2	3	4	5	6	8	10	11	14	16	17			Parzen	Leave-one-out	13.9230		0.0598

(d)

FEATURE VECTOR														CLASSIFIER	ERROR ESTIMATION METHOD	AVERAGE ERROR RATE (%)	K VALUE	h VALUE
2	3	5	6	7	9	10	11	13	15	16	17	18		Bayesian	Resubstitution	19.5305	*	
1	2	3	4	5	6	8	9	12	13	16	17	18		Bayesian	Holdout	7.6389	*	
3	4	5	6	7	8	9	10	12	14	15	16	17		KNN	Holdout	9.0278	8	
1	2	3	4	5	6	7	8	10	11	12	13	14		KNN	Leave-one-out	12.9573	1	
2	3	4	5	6	8	9	10	11	12	14	15	16		Parzen	Holdout	9.0278		0.0798
1	2	3	4	6	7	8	10	11	12	14	16	18		Parzen	Leave-one-out	14.6239		0.0673
2	3	5	6	7	8	9	10	11	13	15	16	17	18	Bayesian	Resubstitution	20.4695	*	
1	2	3	4	5	6	7	8	9	10	12	13	16	17	Bayesian	Holdout	9.2014	*	
1	3	4	5	6	7	8	10	11	12	14	15	16	17	KNN	Holdout	11.8056	1	
1	2	3	4	5	6	7	8	10	11	12	13	14	16	KNN	Leave-one-out	13.2043	1	
1	3	4	5	6	7	8	10	11	12	13	16	17	18	Parzen	Holdout	10.5903		0.0631
1	2	3	4	5	6	7	8	10	11	12	13	14	16	Parzen	Leave-one-out	14.3903		0.0631
2	3	4	5	6	7	9	10	11	12	13	15	16	17	18	Bayesian	Resubstitution	21.5977	*	
1	2	3	4	5	6	7	8	9	10	12	13	16	17	18	Bayesian	Holdout	9.2014	*	
1	3	4	5	6	7	8	10	11	12	13	14	15	16	18	KNN	Holdout	11.9792	1	
1	2	3	4	5	6	7	8	10	11	12	13	14	16	18	KNN	Leave-one-out	14.6239	1	
2	3	4	5	6	7	8	9	11	12	13	14	15	16	17	Parzen	Holdout	10.4167		0.0929
1	2	3	4	5	6	7	9	10	11	12	14	15	16	18	Parzen	Leave-one-out	15.8011		0.0812
1	2	3	4	5	6	7	9	10	11	12	13	15	16	17	18	Bayesian	Resubstitution	22.2897	*	
1	2	3	4	5	6	7	8	9	10	11	13	15	16	17	18	Bayesian	Holdout	10.9375	*	
1	3	4	5	6	7	8	10	11	12	13	14	15	16	17	18	KNN	Holdout	13.7153	1	
1	2	3	4	5	6	7	8	10	11	12	13	15	16	17	18	KNN	Leave-one-out	14.6328	1	
2	3	4	5	6	7	8	9	10	11	12	13	14	15	16	17	Parzen	Holdout	12.1528		0.0877
1	2	3	4	5	6	7	8	9	10	11	12	14	15	16	18	Parzen	Leave-one-out	15.5674		0.0760

(e)

FEATURE VECTOR																		CLASSIFIER	ERROR ESTIMATION METHOD	AVERAGE ERROR RATE (%)	K VALUE	h VALUE
1	2	3	4	5	6	7	8	9	10	11	12	13	15	16	17	18		Bayesian	Resubstitution	24.6484	*	
1	2	3	4	5	6	7	8	9	10	11	12	13	15	16	17	18		Bayesian	Holdout	13.8889	*	
1	2	3	4	6	7	8	9	10	11	12	13	14	15	16	17	18		KNN	Holdout	17.8819	1	
1	2	3	4	5	6	7	8	10	11	12	13	14	15	16	17	18		KNN	Leave-one-out	14.8665	1	
2	3	4	5	6	7	8	9	10	11	12	13	14	15	16	17	18		Parzen	Holdout	13.5417		0.0820
1	2	3	4	5	6	8	9	10	11	12	13	14	15	16	17	18		Parzen	Leave-one-out	16.7490		0.0820
1	2	3	4	5	6	7	8	9	10	11	12	13	14	15	16	17	18	Bayesian	Resubstitution	27.5056	*	
1	2	3	4	5	6	7	8	9	10	11	12	13	14	15	16	17	18	Bayesian	Holdout	18.4028	*	
1	2	3	4	5	6	7	8	9	10	11	12	13	14	15	16	17	18	KNN	Holdout	20.8333	1	
1	2	3	4	5	6	7	8	9	10	11	12	13	14	15	16	17	18	KNN	Leave-one-out	16.5109	1	
1	2	3	4	5	6	7	8	9	10	11	12	13	14	15	16	17	18	Parzen	Holdout	16.4931		0.0981
1	2	3	4	5	6	7	8	9	10	11	12	13	14	15	16	17	18	Parzen	Leave-one-out	18.1375		0.0981

* Bayesian classifier will use the functional form of K in terms of the training data size that was presented by Loftsgaarden and Quesenberry [27]: $K(N_j) = \sqrt{N_j}$ where N_j is the number of labeled observations in the extended object training data set for object class Oj as presented in Table 2.1.

following pairs of classifiers along with their respective error estimation method: (KNN classifier (with holdout method), Parzen classifier (with holdout method)) and (KNN classifier (with leave-one-out method), Parzen classifier (with leave-one-out method)). Within each dimension, the feature vector that is associated with the lowest error rate in each pair of classifiers is selected as a candidate to become a most favorable feature vector. The feature vector with the highest error rate in the pair is eliminated from the set of candidates. Since the Bayesian classifier (with

Table 4.5 Extended object candidates for most favorable feature vectors.

(a)

FEATURE VECTOR							CLASSIFIER	ERROR ESTIMATION METHOD	AVERAGE ERROR RATE (%)	K VALUE	h VALUE
3							Bayesian	Resubstitution	44.3569	*	
6							Bayesian	Holdout	30.0347	*	
6							KNN	Holdout	30.0347	54	
3							Parzen	Leave-one-out	43.9564		0.0123
4	13						Bayesian	Resubstitution	30.0623	*	
6	16						Bayesian	Holdout	13.8889	*	
1	12						KNN	Leave-one-out	30.3894	1	
5	13						Parzen	Holdout	17.8819		0.0433
5	6	18					Bayesian	Resubstitution	23.9697	*	
5	6	16					Bayesian	Holdout	11.9792	*	
1	2	12					KNN	Leave-one-out	10.3427	1	
4	6	16					Parzen	Holdout	9.2014		0.0276
3	5	6	13				Bayesian	Resubstitution	22.0828	*	
5	6	8	16				Bayesian	Holdout	7.2917	*	
1	2	4	14				KNN	Leave-one-out	12.0004	1	
3	6	10	13				Parzen	Holdout	7.8125		0.0381
2	3	5	6	18			Bayesian	Resubstitution	19.0587	*	
3	5	6	9	16			Bayesian	Holdout	5.7292	*	
3	4	7	8	13			KNN	Holdout	6.0764	1	
1	2	3	6	12			KNN	Leave-one-out	10.6030	1	
2	3	6	11	15	18		Bayesian	Resubstitution	18.7939	*	
3	5	8	9	10	17		Bayesian	Holdout	4.3403	*	
1	2	3	6	12	14		KNN	Leave-one-out	10.5986	1	
1	3	5	7	10	16		Parzen	Holdout	5.9028		0.0536

4.5 Model Performance and Feature Selection

(b)

FEATURE VECTOR	CLASSIFIER	ERROR ESTIMATION METHOD	AVERAGE ERROR RATE (%)	K VALUE	h VALUE
2 3 5 6 9 13 18	Bayesian	Resubstitution	16.7045	*	
3 5 6 8 9 13 16	Bayesian	Holdout	2.9514	*	
1 2 4 5 6 10 18	KNN	Leave-one-out	11.5421	1	
1 3 5 6 7 11 16	Parzen	Holdout	4.3403		0.0534
2 3 5 6 9 15 16 18	Bayesian	Resubstitution	16.9737	*	
3 5 6 7 8 9 13 16	Bayesian	Holdout	2.9514	*	
1 3 4 6 11 12 14 18	KNN	Holdout	4.5139	1	
1 2 3 4 5 6 10 18	KNN	Leave-one-out	11.0703	1	
2 3 5 6 7 9 13 15 18	Bayesian	Resubstitution	16.9426	*	
2 3 5 6 7 8 9 11 12	Bayesian	Holdout	2.9514	*	
1 3 4 6 8 11 12 14 18	KNN	Holdout	3.1250	1	
1 2 3 4 5 6 10 14 18	KNN	Leave-one-out	11.3040	1	
2 3 5 6 9 11 13 15 17 18	Bayesian	Resubstitution	17.6747	*	
2 3 5 6 8 9 13 16 17 18	Bayesian	Holdout	4.5139	*	
1 3 4 5 6 8 12 14 16 18	KNN	Holdout	6.0764	1	
1 2 3 4 5 6 10 12 14 18	KNN	Leave-one-out	11.5376	1	
2 3 5 6 7 9 10 13 15 16 17	Bayesian	Resubstitution	17.1495	*	
1 2 3 5 6 8 9 10 13 16 17	Bayesian	Holdout	4.6875	*	
1 3 6 7 8 10 11 12 13 14 18	KNN	Holdout	6.0764	1	
1 2 3 4 5 6 8 10 12 14 18	KNN	Leave-one-out	12.0004	1	

* Bayesian classifier will use the functional form of K in terms of the training data size that was presented by Loftsgaarden and Quesenberry [27]: $K(N_j) = \sqrt{N_j}$ where N_j is the number of labeled observations in the extended object training data set for object class Oj as presented in Table 2.1.

holdout method) and Bayesian classifier (with resubstitution method) both present some uniqueness in the distribution of their error rate trends in Fig. 4.11, all their feature vectors from Table 4.4 will remain as candidates for most favorable feature vectors. The candidates for the most favorable feature vectors are presented in Table 4.5 (a–b). We can now choose a set of most favorable feature vectors that are associated with the lowest error rates within each category of classification models in Table 4.5. A set of most favorable feature vectors is displayed in Table 4.6.

Table 4.6 Extended object set of most favorable feature vectors for each classifier with the respective error estimation method.

FEATURE VECTOR	CLASSIFIER	ERROR ESTIMATION METHOD	AVERAGE ERROR RATE (%)	K VALUE	h VALUE
1 2 12	KNN	Leave-one-out	10.3427	1	
1 2 3 6 12	KNN	Leave-one-out	10.6030	1	
1 2 3 6 12 14	KNN	Leave-one-out	10.5986	1	
2 3 5 6 9 13 18	Bayesian	Resubstitution	16.7045	*	
3 5 6 8 9 13 16	Bayesian	Holdout	2.9514	*	
1 3 5 6 7 11 16	Parzen	Holdout	4.3403		0.0534
2 3 5 6 9 15 16 18	Bayesian	Resubstitution	16.9737	*	
3 5 6 7 8 9 13 16	Bayesian	Holdout	2.9514	*	
1 3 4 6 11 12 14 18	KNN	Holdout	4.5139	1	
2 3 5 6 7 9 13 15 18	Bayesian	Resubstitution	16.9426	*	
2 3 5 6 7 8 9 11 12	Bayesian	Holdout	2.9514	*	
1 3 4 6 8 11 12 14 18	KNN	Holdout	3.1250	1	

* Bayesian classifier will use the functional form of K in terms of the training data size that was presented by Loftsgaarden and Quesenberry [27]: $K(N_j) = \sqrt{N_j}$ where N_j is the number of labeled observations in the extended object training data set for object class Oj as presented in Table 2.1.

Alternatively, we can identify a set of feature vectors that result in minimal error rates for a single classifier on more generalized validation data. As we saw in Sect. 4.5.3, our error estimation methods (resubstitution, holdout, and leave-one-out) choose the test data in different ways. For instance, in the holdout method the training and test data sets are disjoint. On the other hand, in the resubstitution method all the available data used for the training data set is also used as test data. Thus, the performance of the classifier along with a given feature vector is assessed on the test set associated with given error estimation method. By identifying a set of feature vectors in each dimension that simultaneously minimize the error rates on two types of test data sets, we can present a classification model that will provide enough flexibility to ensure acceptable performance on a more generalized test (or blind) data set. The scheme proceeds by first computing the average error rates for a single classifier using two types of error estimation methods for each dimension of features. For each dimension of feature vectors, we will create a scatter plot consisting of the average error rates produced by the single classifier on the two error estimation methods. We will use the scatter plots to determine the feature vector in each dimension that minimize both the average error rates and absolute difference between the average error rates for the single classifier on the two error estimation methods. For example, suppose we consider the combination consisting of the KNN classifier (with the holdout error estimation method) and KNN classifier (with leave-one-out error estimation method) in three dimensions. This combination involves the KNN classifier evaluated on two different test sets determined by their respective error estimation methods. A scatter plot of the average error rates (%) involving the KNN classifier and both of these error estimation methods is displayed in Fig. 4.12. Feature vector < 1, 6, 18 > results in the minimum average error rates with the smallest absolute difference in the error rates on the test data set for each error estimation method used by the KNN classifier. The combination of classifiers and error estimation methods considered in this analysis are: (KNN classifier (with holdout method), KNN classifier (with leave-one-out method)) and (Parzen classifier (with holdout method), Parzen classifier (with leave-one-out method)) and (Bayesian classifier (with holdout method), Bayesian classifier (with resubstitution method)). Table 4.7 (a–c) presents the minimum average error rates with the smallest absolute difference in the error rates on the test data set for each combination across each dimension. After identifying the minimum average error rates in each dimension and combination, we can compare the results and select the most favorable feature vectors associated with the lowest error rates. Once again, the size of the most favorable set of feature vectors is limited to 11 dimensions to support the rule of thumb for peak performance. Table 4.8 displays a set of most favorable feature vectors for the combinations of a classifier and error estimation methods.

Table 4.9 combines the results from Tables 4.6 and 4.8 to present our set of most favorable feature vectors for the extended objects. An important observation is that none of the most favorable feature vectors consist of only features from a single feature category – meteorological, micro, or macro. Additionally, we are choosing a most favorable set of feature vectors rather than identifying the feature vector associated with the overall lowest error rate as the single most favorable feature vector.

4.5 Model Performance and Feature Selection

Table 4.7 Extended object comparison of the lowest average error rates (%) for combinations of a classifier and error estimation methods across each feature vector dimension.

(a)

FEATURE VECTOR	CLASSIFIER	ERROR ESTIMATION METHOD	AVERAGE ERROR RATES (%)		K VALUE	h VALUE
13	KNN	(Holdout, Leave-one-out)	48.0903	49.8709	38	
15	Parzen	(Holdout, Leave-one-out)	62.5000	63.3801		0.0491
14	Bayesian	(Holdout, Resubstitution)	46.3542	48.2933	*	
6 12	KNN	(Holdout, Leave-one-out)	36.2847	36.5042	3	
2 3	Parzen	(Holdout, Leave-one-out)	40.2778	39.8509		0.0136
6 12	Bayesian	(Holdout, Resubstitution)	34.7222	34.3703	*	
1 6 18	KNN	(Holdout, Leave-one-out)	27.2569	26.1571	1	
1 4 13	Parzen	(Holdout, Leave-one-out)	25.8681	25.6809		0.0247
4 7 17	Bayesian	(Holdout, Resubstitution)	27.6042	28.2065	*	
1 5 6 18	KNN	(Holdout, Leave-one-out)	18.0556	18.3756	1	
2 6 11 16	Parzen	(Holdout, Leave-one-out)	19.6181	18.8607		0.0423
2 7 10 13	Bayesian	(Holdout, Resubstitution)	26.2153	25.4295	*	
1 5 6 10 18	KNN	(Holdout, Leave-one-out)	15.1042	15.5585	1	
2 6 7 11 18	Parzen	(Holdout, Leave-one-out)	16.8403	16.2506		0.0491
2 4 5 17 18	Bayesian	(Holdout, Resubstitution)	22.5694	23.3000	*	
1 6 8 10 11 13	KNN	(Holdout, Leave-one-out)	13.7153	14.1522	1	
1 2 3 6 11 16	Parzen	(Holdout, Leave-one-out)	14.9306	15.5763		0.0583
3 9 10 11 15 18	Bayesian	(Holdout, Resubstitution)	21.1806	21.6555	*	
1 6 8 10 11 14 18	KNN	(Holdout, Leave-one-out)	13.8889	14.6061	1	
1 2 3 5 6 13 17	Parzen	(Holdout, Leave-one-out)	15.1042	14.3858		0.0496
2 3 5 10 11 13 17	Bayesian	(Holdout, Resubstitution)	21.0069	21.4486	*	
1 6 8 10 11 12 13 14	KNN	(Holdout, Leave-one-out)	15.2778	14.3769	1	
1 2 3 5 10 11 12 16	Parzen	(Holdout, Leave-one-out)	14.9306	14.6506		0.0646
2 3 5 6 11 13 15 17	Bayesian	(Holdout, Resubstitution)	18.2292	18.3712	*	

(b)

FEATURE VECTOR	CLASSIFIER	ERROR ESTIMATION METHOD	AVERAGE ERROR RATES (%)		K VALUE	h VALUE
1 2 3 4 6 11 13 14 16	KNN	(Holdout, Leave-one-out)	13.5417	14.3903	1	
1 2 3 6 10 11 13 14 16	Parzen	(Holdout, Leave-one-out)	13.5417	13.1954		0.0463
2 3 5 9 10 11 13 15 17	Bayesian	(Holdout, Resubstitution)	18.0556	17.9083	*	
1 2 3 5 6 10 12 13 14 16	KNN	(Holdout, Leave-one-out)	13.5417	13.6805	1	
1 2 3 5 6 7 10 11 14 16	Parzen	(Holdout, Leave-one-out)	13.5417	14.1656		0.0687
2 3 5 9 10 11 13 15 17 18	Bayesian	(Holdout, Resubstitution)	18.2292	18.3845	*	
1 2 3 4 5 6 8 12 13 14 16	KNN	(Holdout, Leave-one-out)	13.5417	13.6760	1	
1 2 3 4 5 6 7 10 12 14 16	Parzen	(Holdout, Leave-one-out)	13.5417	14.1611		0.0637
2 3 5 7 9 10 11 13 15 16 18	Bayesian	(Holdout, Resubstitution)	19.6181	19.8175	*	
1 2 3 5 6 7 8 11 12 14 16 18	KNN	(Holdout, Leave-one-out)	14.9306	15.1046	1	
1 2 4 6 7 8 11 12 14 16 17	Parzen	(Holdout, Leave-one-out)	15.1042	15.1046		0.0715
2 3 5 7 9 10 11 13 15 16 17 18	Bayesian	(Holdout, Resubstitution)	19.6181	20.0512	*	
1 2 3 4 6 7 8 10 11 12 14 16 17	KNN	(Holdout, Leave-one-out)	14.9306	14.3947	1	
1 2 3 4 5 6 7 8 10 11 12 14 16	Parzen	(Holdout, Leave-one-out)	15.1042	15.0957		0.0735
1 2 3 5 7 9 10 11 13 15 16 17 18	Bayesian	(Holdout, Resubstitution)	22.3958	22.3787	*	
1 2 3 4 5 6 7 8 11 12 13 14 16 18	KNN	(Holdout, Leave-one-out)	15.1042	15.0957	1	
1 2 3 4 5 6 7 10 11 12 14 15 16 18	Parzen	(Holdout, Leave-one-out)	16.4931	16.5198		0.0744
2 4 5 6 7 9 11 12 13 14 15 16 17 18	Bayesian	(Holdout, Resubstitution)	24.1319	23.9964	*	
1 2 3 5 6 7 8 11 12 13 14 15 16 17 18	KNN	(Holdout, Leave-one-out)	16.4931	16.2906	1	
1 2 3 4 5 6 7 9 10 11 12 13 14 15 16	Parzen	(Holdout, Leave-one-out)	16.4931	16.2639		0.0869
1 2 4 5 6 7 9 11 12 13 14 15 16 17 18	Bayesian	(Holdout, Resubstitution)	27.2569	28.6871	*	
1 2 3 4 6 7 9 10 11 12 13 14 15 16 17 18	KNN	(Holdout, Leave-one-out)	17.8819	17.9261	1	
1 2 3 4 5 7 8 9 10 11 12 13 14 15 17 18	Parzen	(Holdout, Leave-one-out)	18.0556	18.3890		0.0930
1 2 3 4 5 6 7 9 11 12 13 14 15 16 17 18	Bayesian	(Holdout, Resubstitution)	22.7431	26.3373	*	

(c)

FEATURE VECTOR	CLASSIFIER	ERROR ESTIMATION METHOD	AVERAGE ERROR RATES (%)	K VALUE	h VALUE
1 2 3 4 5 6 7 8 9 10 11 12 13 14 15 16 17 18	KNN	(Holdout, Leave-one-out)	17.8819 17.2074	1	
1 2 3 4 5 6 7 8 9 10 11 12 13 14 15 16 17 18	Parzen	(Holdout, Leave-one-out)	18.0556 18.6093		0.0963
1 2 3 4 5 6 7 8 9 10 11 12 13 14 15 16 17 18	Bayesian	(Holdout, Resubstitution)	19.9653 27.0427	*	
1 2 3 4 5 6 7 8 9 10 11 12 13 14 15 16 17 18	KNN	(Holdout, Leave-one-out)	20.8333 16.5109	1	
1 2 3 4 5 6 7 8 9 10 11 12 13 14 15 16 17 18	Parzen	(Holdout, Leave-one-out)	16.4931 18.1375		0.0981
1 2 3 4 5 6 7 8 9 10 11 12 13 14 15 16 17 18	Bayesian	(Holdout, Resubstitution)	18.4028 27.5056	*	

* Bayesian classifier will use the functional form of K in terms of the training data size that was presented by Loftsgaarden and Quesenberry [27]: $K(N_j) = \sqrt{N_j}$ where N_j is the number of labeled observations in the extended object training data set for object class Oj as presented in Table 2.1.

Table 4.8 Extended object set of most favorable feature vectors for combinations of a classifier and error estimation methods.

FEATURE VECTOR	CLASSIFIER	ERROR ESTIMATION METHOD	AVERAGE ERROR RATES (%)	K VALUE	h VALUE
1 2 3 4 6 11 13 14 16	KNN	(Holdout, Leave-one-out)	13.5417 14.3903	1	
1 2 3 6 10 11 13 14 16	Parzen	(Holdout, Leave-one-out)	13.5417 13.1954		0.0463
2 3 5 9 10 11 13 15 17	Bayesian	(Holdout, Resubstitution)	18.0556 17.9083	*	
1 2 3 5 6 10 12 13 14 16	KNN	(Holdout, Leave-one-out)	13.5417 13.6805	1	
1 2 3 5 6 7 10 11 14 16	Parzen	(Holdout, Leave-one-out)	13.5417 14.1656		0.0687
1 2 3 4 5 6 8 12 13 14 16	KNN	(Holdout, Leave-one-out)	13.5417 13.6760	1	
1 2 3 4 5 6 7 10 12 14 16	Parzen	(Holdout, Leave-one-out)	13.5417 14.1611		0.0637

* Bayesian classifier will use the functional form of K in terms of the training data size that was presented by Loftsgaarden and Quesenberry [27]: $K(N_j) = \sqrt{N_j}$ where N_j is the number of labeled observations in the extended object training data set for object class Oj as presented in Table 2.1.

Table 4.9 Extended object set of most favorable feature vectors (combined feature vectors from Tables 4.6 and 4.8).

FEATURE VECTOR	CLASSIFIER	ERROR ESTIMATION METHOD	AVERAGE ERROR RATES (%)	K VALUE	h VALUE
1 2 12	KNN	Leave-one-out	10.3427	1	
1 2 3 6 12	KNN	Leave-one-out	10.6030	1	
1 2 3 6 12 14	KNN	Leave-one-out	10.5986	1	
2 3 5 6 9 13 18	Bayesian	Resubstitution	16.7045	*	
3 5 6 8 9 13 16	Bayesian	Holdout	2.9514	*	
1 3 5 6 7 11 16	Parzen	Holdout	4.3403		0.0534
2 3 5 6 9 15 16 18	Bayesian	Resubstitution	16.9737	*	
3 5 6 7 8 9 13 16	Bayesian	Holdout	2.9514	*	
1 3 4 6 11 12 14 18	KNN	Holdout	4.5139	1	
2 3 5 6 7 9 13 15 18	Bayesian	Resubstitution	16.9426	*	
2 3 5 6 7 8 9 11 12	Bayesian	Holdout	2.9514	*	
1 3 4 6 8 11 12 14 18	KNN	Holdout	3.1250	1	
1 2 3 4 6 11 13 14 16	KNN	(Holdout, Leave-one-out)	13.5417 14.3903	1	
1 2 3 6 10 11 13 14 16	Parzen	(Holdout, Leave-one-out)	13.5417 13.1954		0.0463
2 3 5 9 10 11 13 15 17	Bayesian	(Holdout, Resubstitution)	18.0556 17.9083	*	
1 2 3 5 6 10 12 13 14 16	KNN	(Holdout, Leave-one-out)	13.5417 13.6805	1	
1 2 3 5 6 7 10 11 14 16	Parzen	(Holdout, Leave-one-out)	13.5417 14.1656		0.0687
1 2 3 4 5 6 8 12 13 14 16	KNN	(Holdout, Leave-one-out)	13.5417 13.6760	1	
1 2 3 4 5 6 7 10 12 14 16	Parzen	(Holdout, Leave-one-out)	13.5417 14.1611		0.0637

* Bayesian classifier will use the functional form of K in terms of the training data size that was presented by Loftsgaarden and Quesenberry [27]: $K(N_j) = \sqrt{N_j}$ where N_j is the number of labeled observations in the extended object training data set for object class Oj as presented in Table 2.1.

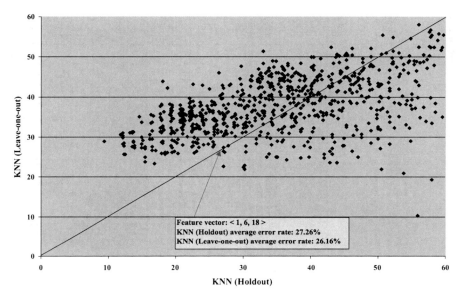

Fig. 4.12 Extended object scatter plot of average error rates (%) for KNN classifier (with holdout error estimation method) and KNN classifier (with leave-one-out error estimation method) in three dimensions. Feature vector < 1, 6, 18 > results in the minimum average error rates with the smallest absolute difference in the error rates on the test data set for each error estimation method used by the KNN classifier.

Considering a set of most favorable feature vectors will allow us to design a classification model that is able to generalize to other test (or blind) data sets, rather than choosing a single feature vector that results in a model with too little flexibility.

4.5.6 Compact Object Performance and Feature Selection

We will now repeat the same procedures presented in Sect. 4.5.5 to identify the most favorable features for the compact objects. The 15 thermal features remaining from our preliminary feature analysis in Sect. 4.3 are displayed in Table 4.10 along with numerical labels that are provided for convenience as we analyze the different feature vectors used in the classification models during the exhaustive search feature selection method. The equations for each feature were discussed and derived in Chap. 3.

Table 4.11 provides the number of combinations of extended object features for each feature vector dimension used in the exhaustive search method. We will compute the average error rates for the classification models across all 15 dimensions in our exhaustive search. Since the number of training patterns for each of the compact object classes (steel pole and tree) is $n = 318$ as displayed in Table 2.1, the rule of thumb for peak performance given in Sect. 4.5.1 limits us to a feature vector

Table 4.10 Compact object thermal features and labels used in the exhaustive search feature selection method.

FEATURE	LABEL
Meteorological Features	
Ambient Temp. $°F$ (Ta)	1
Amb. Temp. Rate of Change (T1)	2
Micro Features	
Object Surface Radiance (Lo)	3
Reference Emitter Radiance (Lr)	4
Background Irradiance (Lb)	5
Lo/Lr (Lor)	6
Lo/Lob (Lob)	7
Emissivity (Eo)	8
Macro Features	
First-order Statistics	
Contrast1 (Co1)	9
Entropy1 (En1)	10
Second-order Statistics	
Contrast2 (Co2)	11
Correlation (Cr2)	12
Energy (Er2)	13
Homogeneity (Ho2)	14
Entropy2 (En2)	15

with up to 32 features. Thus, we could consider all 15 dimensions in our analysis to identify a set of most favorable feature vectors for our compact objects.

The average error rates for each classifier combined with every combination of features across all 15 dimensions (i.e., exhaustive search feature selection method) were computed using the error estimation methods on a high performance computing system discussed in Sect. 4.5. Figure 4.13 presents dotplots that give the general trend of the average error rates for each classifier and error estimation method observed in each dimension. Similar to the extended objects, the dotplots for the compact objects show that holdout error estimation methods have a tendency to display a higher variance in the average error rates compared to the leave-one-out and resubstitution methods. Once again, this result occurs since the training and test data come from the same set for the leave-one-out and resubstitution methods.

The average error rates were sorted in increasing order by classifier and error estimation method within each dimension. Tables 4.12 (a–d) compares the lowest average error rates (%) of each classifier with the respective error estimation method across each feature vector dimension. The average error rates in Tables 4.12 (a–d) display the behavior of the peaking phenomenon. As the number of dimension of a feature vector increases, the error rates of each classifier decrease to a specific peak (or in some cases a short plateau) and then the classification performance begins to decline. For instance, the Bayesian classifier with the resubstitution error estimation

4.5 Model Performance and Feature Selection

Table 4.11 Total number of compact object thermal feature combinations for feature vectors from 1 to 15 dimensions. All 15 dimensions satisfy the rule of thumb to ensure peak performance of the classification models.

DIMENSIONS OF FEATURE VECTOR	NUMBER OF FEATURE COMBINATIONS
1	15
2	105
3	455
4	1365
5	3003
6	5005
7	6435
8	6435
9	5005
10	3003
11	1365
12	455
13	105
14	15
15	1
TOTAL	**32767**

method reaches its peak performance at an estimated average error rate of 6.45% with a 7-dimensional feature vector and maintains this error rate up to nine dimensions before the performance begins to decline. We also see that no classifier reaches a peak performance with a feature vector consisting of only features from a single feature category – meteorological, micro, or macro.

The next step is to compare classification models (along with their respective error estimation method) in Tables 4.12 (a–d) to identify a most favorable set of feature vectors. We will choose pairs of classification models for comparison based on their similarities in the error rate trends found in Fig. 4.13. Thus, for each dimension, we will compare the error rates in Table 4.12 for the following pairs of classifiers along with their respective error estimation method: (KNN classifier (with holdout method) method), Parzen classifier (with holdout method)) and (KNN classifier (with leave-one-out method), Parzen classifier (with leave-one-out method)). Within each dimension, the feature vector that is associated with the lowest error rate in each pair of classifiers is selected as a candidate to become a most favorable feature vector. The feature vector with the highest error rate in the pair is eliminated from the set of candidates. Since the Bayesian classifier (with holdout method) and Bayesian classifier (with resubstitution method) both present some uniqueness in the distribution of their error rate trends in Fig. 4.13, all their feature vectors from Table 4.12 will remain as candidates for most favorable feature vectors. The candidates for the most favorable feature vectors are presented in Table 4.13 (a–c). We retained both feature vectors in any pairs that had equal error rates. We can now choose a set of most favorable feature vectors that are associated with the lowest error rates within each category of classification models in Table 4.13. A set of most favorable feature vectors is displayed in Table 4.14.

136 4 Thermal Feature Selection

Table 4.12 Compact object comparison of the lowest average error rates (%) of each classifier with the respective error estimation method across each feature vector dimension.

(a)

FEATURE VECTOR	CLASSIFIER	ERROR ESTIMATION METHOD	AVERAGE ERROR RATE (%)	K VALUE	h VALUE
8	Bayesian	Resubstitution	11.9497	*	
8	Bayesian	Holdout	8.3333	*	
8	KNN	Holdout	8.2532	19	
8	KNN	Leave-one-out	11.6352	19	
8	Parzen	Holdout	8.2532		0.03642
8	Parzen	Leave-one-out	12.7358		0.03642
7 14	Bayesian	Resubstitution	8.6478	*	
2 8	Bayesian	Holdout	7.2917	*	
6 8	KNN	Holdout	6.2500	9	
7 14	KNN	Leave-one-out	8.3333	25	
8 11	Parzen	Holdout	8.2532		0.00869
7 14	Parzen	Leave-one-out	8.8050		0.02764
4 7 14	Bayesian	Resubstitution	7.5472	*	
1 8 14	Bayesian	Holdout	5.1282	*	
2 6 8	KNN	Holdout	6.1699	8	
4 7 14	KNN	Leave-one-out	7.2327	34	
4 8 14	Parzen	Holdout	5.9295		0.04974
4 7 14	Parzen	Leave-one-out	7.7044		0.0343
6 7 9 14	Bayesian	Resubstitution	7.0755	*	
6 8 13 14	Bayesian	Holdout	2.9647	*	
6 7 10 14	KNN	Holdout	4.0865	6	
4 7 11 14	KNN	Leave-one-out	7.0755	10	
3 6 8 14	Parzen	Holdout	4.0865		0.05049
4 7 9 14	Parzen	Leave-one-out	7.2327		0.03786

(b)

FEATURE VECTOR	CLASSIFIER	ERROR ESTIMATION METHOD	AVERAGE ERROR RATE (%)	K VALUE	h VALUE
1 3 6 7 14	Bayesian	Resubstitution	6.7610	*	
3 6 8 11 13	Bayesian	Holdout	2.9647	*	
6 7 10 11 14	KNN	Holdout	4.0865	6	
1 6 7 10 14	KNN	Leave-one-out	6.6038	7	
3 6 8 9 14	Parzen	Holdout	4.0064		0.06232
1 2 7 8 14	Parzen	Leave-one-out	7.2327		0.05171
1 3 6 7 12 14	Bayesian	Resubstitution	6.4465	*	
4 6 8 9 11 13	Bayesian	Holdout	2.0032	*	
6 7 9 10 11 14	KNN	Holdout	4.0865	6	
1 5 6 7 12 15	KNN	Leave-one-out	6.1321	7	
3 6 8 9 11 14	Parzen	Holdout	3.0449		0.06099
1 2 6 7 11 15	Parzen	Leave-one-out	7.3899		0.04437
1 2 6 7 9 10 14	Bayesian	Resubstitution	6.4465	*	
1 6 7 11 12 13 15	Bayesian	Holdout	2.0032	*	
3 6 8 9 10 11 14	KNN	Holdout	4.0064	15	
1 5 6 7 10 12 14	KNN	Leave-one-out	6.1321	5	
1 4 6 8 9 14 15	Parzen	Holdout	4.0064		0.07544
1 5 6 7 11 12 15	Parzen	Leave-one-out	7.0755		0.05542
1 2 4 6 7 9 13 14	Bayesian	Resubstitution	6.4465	*	
1 6 7 9 11 12 13 14	Bayesian	Holdout	2.0032	*	
2 3 6 8 9 11 13 15	KNN	Holdout	4.0865	14	
1 5 6 7 9 10 12 14	KNN	Leave-one-out	5.8176	5	
2 6 7 8 9 10 11 12	Parzen	Holdout	4.1667		0.06702
1 5 6 7 9 10 12 14	Parzen	Leave-one-out	7.0755		0.06474

4.5 Model Performance and Feature Selection

(c)

FEATURE VECTOR	CLASSIFIER	ERROR ESTIMATION METHOD	AVERAGE ERROR RATE (%)	K VALUE	h VALUE
1 2 4 6 7 9 10 13 14	Bayesian	Resubstitution	6.4465	*	
1 6 7 8 10 11 12 13 14	Bayesian	Holdout	2.0032	*	
2 3 4 8 9 10 11 13 15	KNN	Holdout	4.0865	14	
1 5 6 7 9 10 11 12 14	KNN	Leave-one-out	5.8176	5	
6 7 8 9 10 12 13 14 15	Parzen	Holdout	4.0865		0.07706
1 5 6 7 9 10 11 12 14	Parzen	Leave-one-out	7.0755		0.06533
1 3 4 6 7 9 10 13 14 15	Bayesian	Resubstitution	6.7610	*	
1 6 7 9 10 11 12 13 14 15	Bayesian	Holdout	2.0032	*	
1 3 4 6 8 9 11 13 14 15	KNN	Holdout	4.0865	14	
1 4 5 6 7 9 10 11 12 14	KNN	Leave-one-out	6.4465	7	
2 3 6 7 8 9 10 11 12 15	Parzen	Holdout	4.1667		0.07228
1 3 5 6 7 8 9 10 11 14	Parzen	Leave-one-out	7.2327		0.08107
1 2 3 4 5 6 7 8 9 10 14	Bayesian	Resubstitution	7.3899	*	
1 2 4 6 7 8 10 11 13 14 15	Bayesian	Holdout	2.0032	*	
1 2 4 6 7 8 9 11 13 14 15	KNN	Holdout	4.1667	7	
1 3 4 5 6 7 9 10 11 12 14	KNN	Leave-one-out	6.6038	7	
2 3 6 7 8 9 10 11 12 14 15	Parzen	Holdout	5.1282		0.07677
1 2 5 6 7 8 9 11 12 14 15	Parzen	Leave-one-out	7.2327		0.08712
1 3 4 5 6 7 8 9 10 12 13 14	Bayesian	Resubstitution	7.5472	*	
1 2 4 6 7 8 9 11 12 13 14 15	Bayesian	Holdout	2.9647	*	
2 3 4 6 7 8 9 10 11 13 14 15	KNN	Holdout	5.1282	5	
1 3 4 5 6 7 9 10 11 12 14 15	KNN	Leave-one-out	6.7610	7	
1 2 3 6 8 9 10 11 12 13 14 15	Parzen	Holdout	7.1314		0.09115
1 2 3 4 5 6 7 8 9 11 12 14	Parzen	Leave-one-out	7.2327		0.07915

(d)

FEATURE VECTOR	CLASSIFIER	ERROR ESTIMATION METHOD	AVERAGE ERROR RATE (%)	K VALUE	h VALUE
1 2 3 5 6 7 8 9 10 12 13 14 15	Bayesian	Resubstitution	8.0189	*	
1 2 4 6 7 8 9 10 11 12 13 14 15	Bayesian	Holdout	4.0064	*	
1 2 3 4 6 7 8 9 10 11 13 14 15	KNN	Holdout	6.2500	7	
1 2 3 4 5 6 7 8 9 10 11 14 15	KNN	Leave-one-out	7.0755	11	
2 3 4 5 6 8 9 10 11 12 13 14 15	Parzen	Holdout	8.2532		0.08932
1 2 3 4 5 6 7 8 10 11 12 14 15	Parzen	Leave-one-out	7.7044		0.09106
1 2 3 4 5 6 7 8 9 10 12 13 14 15	Bayesian	Resubstitution	8.0189	*	
1 2 3 4 5 6 7 9 10 11 12 13 14 15	Bayesian	Holdout	5.1282	*	
1 2 3 4 6 7 8 9 10 11 12 13 14 15	KNN	Holdout	6.2500	7	
1 2 3 4 5 6 7 8 9 10 11 12 14 15	KNN	Leave-one-out	7.3899	7	
1 2 3 4 5 6 8 9 10 11 12 13 14 15	Parzen	Holdout	8.3333		0.0953
1 2 3 4 5 6 7 8 9 10 11 12 14 15	Parzen	Leave-one-out	8.0189		0.08984
1 2 3 4 5 6 7 8 9 10 11 12 13 14 15	Bayesian	Resubstitution	10.5346	*	
1 2 3 4 5 6 7 8 9 10 11 12 13 14 15	Bayesian	Holdout	7.2115	*	
1 2 3 4 5 6 7 8 9 10 11 12 13 14 15	KNN	Holdout	9.3750	7	
1 2 3 4 5 6 7 8 9 10 11 12 13 14 15	KNN	Leave-one-out	8.6478	7	
1 2 3 4 5 6 7 8 9 10 11 12 13 14 15	Parzen	Holdout	9.3750		0.09889
1 2 3 4 5 6 7 8 9 10 11 12 13 14 15	Parzen	Leave-one-out	8.8050		0.09889

* Bayesian classifier will use the functional form of K in terms of the training data size that was presented by Loftsgaarden and Quesenberry [27]: $K(N_j) = \sqrt{N_j}$ where N_j is the number of labeled observations in the extended object training data set for object class Oj as presented in Table 2.1.

4 Thermal Feature Selection

Table 4.13 Compact object candidates for most favorable feature vectors.

(a)

FEATURE VECTOR	CLASSIFIER	ERROR ESTIMATION METHOD	AVERAGE ERROR RATE (%)	K VALUE	h VALUE
8	Bayesian	Resubstitution	11.9497	*	
8	Bayesian	Holdout	8.3333	*	
8	KNN	Holdout	8.2532	19	
8	KNN	Leave-one-out	11.6352	19	
8	Parzen	Holdout	8.2532		0.03642
7 14	Bayesian	Resubstitution	8.6478	*	
2 8	Bayesian	Holdout	7.2917	*	
6 8	KNN	Holdout	6.2500	9	
7 14	KNN	Leave-one-out	8.3333	25	
4 7 14	Bayesian	Resubstitution	7.5472	*	
1 8 14	Bayesian	Holdout	5.1282	*	
4 7 14	KNN	Leave-one-out	7.2327	34	
4 8 14	Parzen	Holdout	5.9295		0.04974
6 7 9 14	Bayesian	Resubstitution	7.0755	*	
6 8 13 14	Bayesian	Holdout	2.9647	*	
6 7 10 14	KNN	Holdout	4.0865	6	
4 7 11 14	KNN	Leave-one-out	7.0755	10	
3 6 8 14	Parzen	Holdout	4.0865		0.05049
1 3 6 7 14	Bayesian	Resubstitution	6.7610	*	
3 6 8 11 13	Bayesian	Holdout	2.9647	*	
1 6 7 10 14	KNN	Leave-one-out	6.6038	7	
3 6 8 9 14	Parzen	Holdout	4.0064		0.06232

(b)

FEATURE VECTOR	CLASSIFIER	ERROR ESTIMATION METHOD	AVERAGE ERROR RATE (%)	K VALUE	h VALUE
1 3 6 7 12 14	Bayesian	Resubstitution	6.4465	*	
4 6 8 9 11 13	Bayesian	Holdout	2.0032	*	
1 5 6 7 12 15	KNN	Leave-one-out	6.1321	7	
3 6 8 9 11 14	Parzen	Holdout	3.0449		0.06099
1 2 6 7 9 10 14	Bayesian	Resubstitution	6.4465	*	
1 6 7 11 12 13 15	Bayesian	Holdout	2.0032	*	
3 6 8 9 10 11 14	KNN	Holdout	4.0064	15	
1 5 6 7 10 12 14	KNN	Leave-one-out	6.1321	5	
1 4 6 8 9 14 15	Parzen	Holdout	4.0064		0.07544
1 2 4 6 7 9 13 14	Bayesian	Resubstitution	6.4465	*	
1 6 7 9 11 12 13 14	Bayesian	Holdout	2.0032	*	
2 3 6 8 9 11 13 15	KNN	Holdout	4.0865	14	
1 5 6 7 9 10 12 14	KNN	Leave-one-out	5.8176	5	
1 2 4 6 7 9 10 13 14	Bayesian	Resubstitution	6.4465	*	
1 6 7 8 10 11 12 13 14	Bayesian	Holdout	2.0032	*	
2 3 4 8 9 10 11 13 15	KNN	Holdout	4.0865	14	
1 5 6 7 9 10 11 12 14	KNN	Leave-one-out	5.8176	5	
6 7 8 9 10 12 13 14 15	Parzen	Holdout	4.0865		0.07706
1 3 4 6 7 9 10 13 14 15	Bayesian	Resubstitution	6.7610	*	
1 6 7 9 10 11 12 13 14 15	Bayesian	Holdout	2.0032	*	
1 3 4 6 8 9 11 13 14 15	KNN	Holdout	4.0865	14	
1 4 5 6 7 9 10 11 12 14	KNN	Leave-one-out	6.4465	7	

(c)

FEATURE VECTOR															CLASSIFIER	ERROR ESTIMATION METHOD	AVERAGE ERROR RATE (%)	K VALUE	h VALUE
1	2	3	4	5	6	7	8	9	10	14					Bayesian	Resubstitution	7.3899	*	
1	2	4	6	7	8	10	11	13	14	15					Bayesian	Holdout	2.0032	*	
1	2	4	6	7	8	9	11	13	14	15					KNN	Holdout	4.1667	7	
1	3	4	5	6	7	9	10	11	12	14					KNN	Leave-one-out	6.6038	7	
1	3	4	5	6	7	8	9	10	12	13	14				Bayesian	Resubstitution	7.5472	*	
1	2	4	6	7	8	9	11	12	13	14	15				Bayesian	Holdout	2.9647	*	
2	3	4	6	7	8	9	10	11	13	14	15				KNN	Holdout	5.1282	5	
1	3	4	5	6	7	9	10	11	12	14	15				KNN	Leave-one-out	6.7610	7	
1	2	3	5	6	7	8	9	10	12	13	14	15			Bayesian	Resubstitution	8.0189	*	
1	2	4	6	7	8	9	10	11	12	13	14	15			Bayesian	Holdout	4.0064	*	
1	2	3	4	6	7	8	9	10	11	13	14	15			KNN	Holdout	6.2500	7	
1	2	3	4	5	6	7	8	9	10	11	14	15			KNN	Leave-one-out	7.0755	11	
1	2	3	4	5	6	7	8	9	10	12	13	14	15		Bayesian	Resubstitution	8.0189	*	
1	2	3	4	5	6	7	9	10	11	12	13	14	15		Bayesian	Holdout	5.1282	*	
1	2	3	4	6	7	8	9	10	11	12	13	14	15		KNN	Holdout	6.2500	7	
1	2	3	4	5	6	7	8	9	10	11	12	14	15		KNN	Leave-one-out	7.3899	7	
1	2	3	4	5	6	7	8	9	10	11	12	13	14	15	Bayesian	Resubstitution	10.5346	*	
1	2	3	4	5	6	7	8	9	10	11	12	13	14	15	Bayesian	Holdout	7.2115	*	
1	2	3	4	5	6	7	8	9	10	11	12	13	14	15	KNN	Holdout	9.3750	7	
1	2	3	4	5	6	7	8	9	10	11	12	13	14	15	KNN	Leave-one-out	8.6478	7	
1	2	3	4	5	6	7	8	9	10	11	12	13	14	15	Parzen	Holdout	9.3750		0.09889

* Bayesian classifier will use the functional form of K in terms of the training data size that was presented by Loftsgaarden and Quesenberry [27]: $K(N_j) = \sqrt{N_j}$ where N_j is the number of labeled observations in the extended object training data set for object class Oj as presented in Table 2.1.

Table 4.14 Compact object set of most favorable feature vectors for each classifier with the respective error estimation method.

FEATURE VECTOR									CLASSIFIER	ERROR ESTIMATION METHOD	AVERAGE ERROR RATE (%)	K VALUE	h VALUE
1	3	6	7	12	14				Bayesian	Resubstitution	6.4465	*	
4	6	8	9	11	13				Bayesian	Holdout	2.0032	*	
3	6	8	9	11	14				Parzen	Holdout	3.0449		0.06099
1	2	6	7	9	10	14			Bayesian	Resubstitution	6.4465	*	
1	6	7	11	12	13	15			Bayesian	Holdout	2.0032	*	
3	6	8	9	10	11	14			KNN	Holdout	4.0064	15	
1	2	4	6	7	9	13	14		Bayesian	Resubstitution	6.4465	*	
1	6	7	9	11	12	13	14		Bayesian	Holdout	2.0032	*	
1	5	6	7	9	10	12	14		KNN	Leave-one-out	5.8176	5	
1	2	4	6	7	9	10	13	14	Bayesian	Resubstitution	6.4465	*	
1	6	7	8	10	11	12	13	14	Bayesian	Holdout	2.0032	*	
1	5	6	7	9	10	11	12	14	KNN	Leave-one-out	5.8176	5	
1	6	7	9	10	11	12	13	14	15	Bayesian	Holdout	2.0032	*	
1	2	4	6	7	8	10	11	13	14	15	Bayesian	Holdout	2.0032	*	

* Bayesian classifier will use the functional form of K in terms of the training data size that was presented by Loftsgaarden and Quesenberry [27]: $K(N_j) = \sqrt{N_j}$ where N_j is the number of labeled observations in the extended object training data set for object class Oj as presented in Table 2.1.

140   4 Thermal Feature Selection

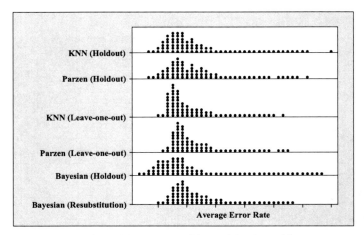

Fig. 4.13 General trend for compact objects of dotplots with average error rates for each classifier and error estimation method observed in each dimension.

As in Sect. 4.5.5, we will now identify a set of most favorable feature vectors involving combinations of a classifier and error estimation methods. The set of most feature vectors are associated with classification models that display flexibility by yielding acceptable performance on a more generalized test (or blind) data set. Following the same scheme in Sect. 4.5.5, we proceed by first computing the average error rates for a single classifier using two types of error estimation methods for each dimension of features. As discussed in the previous section, the two error estimation methods choose their respective test data in different ways. For each dimension of

Table 4.15 Compact object comparison of the lowest average error rates (%) for combinations of a classifier and error estimation methods across each feature vector dimension.

(a)

FEATURE VECTOR											CLASSIFIER	ERROR ESTIMATION METHOD	AVERAGE ERROR RATES (%)		K VALUE	h VALUE
7											KNN	(Holdout, Leave-one-out)	11.4583	12.7358	12	
7											Parzen	(Holdout, Leave-one-out)	13.5417	13.2075		0.011912
7											Bayesian	(Holdout, Resubstitution)	11.4583	12.7358	*	
3	8										KNN	(Holdout, Leave-one-out)	10.4167	10.6918	13	
7	8										Parzen	(Holdout, Leave-one-out)	12.5000	12.2642		0.0379
7	12										Bayesian	(Holdout, Resubstitution)	10.4167	10.6918	*	
6	7	14									KNN	(Holdout, Leave-one-out)	8.2532	8.3333	26	
7	8	14									Parzen	(Holdout, Leave-one-out)	9.3750	8.9623		0.0506
7	8	14									Bayesian	(Holdout, Resubstitution)	9.3750	9.2767	*	
6	7	9	14								KNN	(Holdout, Leave-one-out)	7.2917	7.8616	9	
1	6	7	14								Parzen	(Holdout, Leave-one-out)	7.2917	7.7044		0.0486
6	7	12	14								Bayesian	(Holdout, Resubstitution)	7.2917	7.3899	*	
1	6	7	12	15							KNN	(Holdout, Leave-one-out)	7.2917	7.0755	7	
1	6	7	12	14							Parzen	(Holdout, Leave-one-out)	7.2917	8.1761		0.0517
6	7	9	12	14							Bayesian	(Holdout, Resubstitution)	7.2917	7.5472	*	
4	5	6	7	12	15						KNN	(Holdout, Leave-one-out)	7.2115	7.7044	5	
1	6	7	11	12	15						Parzen	(Holdout, Leave-one-out)	7.2917	8.0189		0.0533
1	6	7	9	10	14						Bayesian	(Holdout, Resubstitution)	6.2500	6.7610	*	

4.5 Model Performance and Feature Selection   141

(b)

FEATURE VECTOR	CLASSIFIER	ERROR ESTIMATION METHOD	AVERAGE ERROR RATES (%)		K VALUE	h VALUE
1 4 6 7 9 10 14	KNN	(Holdout, Leave-one-out)	7.2917	7.3899	6	
1 6 7 8 9 10 14	Parzen	(Holdout, Leave-one-out)	7.2115	7.5472		0.0653
1 2 6 7 10 12 14	Bayesian	(Holdout, Resubstitution)	7.2917	7.5472	*	
1 3 4 5 6 12 14 15	KNN	(Holdout, Leave-one-out)	7.2917	7.5472	7	
1 6 7 8 9 10 11 14	Parzen	(Holdout, Leave-one-out)	7.2115	7.5472		0.0650
3 4 6 7 9 10 14 15	Bayesian	(Holdout, Resubstitution)	6.2500	6.7610	*	
1 3 4 5 6 7 8 10 14	KNN	(Holdout, Leave-one-out)	7.2917	7.3899	6	
1 3 4 6 7 11 12 14 15	Parzen	(Holdout, Leave-one-out)	8.3333	8.8050		0.0643
1 3 6 7 9 10 13 14 15	Bayesian	(Holdout, Resubstitution)	6.2500	6.9182	*	
1 3 4 5 6 7 9 10 11 14	KNN	(Holdout, Leave-one-out)	7.2917	7.0755	6	
1 3 4 6 7 9 11 12 14 15	Parzen	(Holdout, Leave-one-out)	8.3333	8.8050		0.0635
1 2 3 6 7 9 10 13 14 15	Bayesian	(Holdout, Resubstitution)	6.2500	6.9182	*	
1 3 4 5 6 7 8 9 11 14 15	KNN	(Holdout, Leave-one-out)	7.2917	7.7044	6	
1 3 4 6 7 8 9 10 11 12 14	Parzen	(Holdout, Leave-one-out)	8.3333	8.8050		0.0721
1 2 3 4 6 7 8 10 12 13 14	Bayesian	(Holdout, Resubstitution)	7.2917	7.5472	*	
1 2 3 4 5 6 7 9 10 11 14 15	KNN	(Holdout, Leave-one-out)	7.2917	7.5472	5	
1 3 4 5 6 9 10 11 12 13 14 15	Parzen	(Holdout, Leave-one-out)	8.3333	8.9623		0.0761
1 2 4 5 6 7 8 9 10 13 14 15	Bayesian	(Holdout, Resubstitution)	8.3333	8.4906	*	

(c)

FEATURE VECTOR	CLASSIFIER	ERROR ESTIMATION METHOD	AVERAGE ERROR RATES (%)		K VALUE	h VALUE
1 3 4 5 6 7 8 9 10 11 12 14 15	KNN	(Holdout, Leave-one-out)	7.2917	7.7044	3	
3 4 5 6 7 8 9 10 11 12 13 14 15	Parzen	(Holdout, Leave-one-out)	9.3750	9.9057		0.0893
1 2 3 4 5 6 8 9 10 12 13 14 15	Bayesian	(Holdout, Resubstitution)	9.2949	9.7484	*	
2 3 4 5 6 7 8 9 10 11 12 13 14 15	KNN	(Holdout, Leave-one-out)	8.3333	8.4906	11	
1 3 4 5 6 7 8 9 10 11 12 13 14 15	Parzen	(Holdout, Leave-one-out)	9.3750	9.4340		0.0953
1 2 3 4 5 6 7 8 9 10 11 12 13 15	Bayesian	(Holdout, Resubstitution)	9.2949	10.0629	*	
1 2 3 4 5 6 7 8 9 10 11 12 13 14 15	KNN	(Holdout, Leave-one-out)	9.3750	8.6478	7	
1 2 3 4 5 6 7 8 9 10 11 12 13 14 15	Parzen	(Holdout, Leave-one-out)	9.3750	8.8050		0.0989
1 2 3 4 5 6 7 8 9 10 11 12 13 14 15	Bayesian	(Holdout, Resubstitution)	7.2115	10.5346	*	

* Bayesian classifier will use the functional form of K in terms of the training data size that was presented by Loftsgaarden and Quesenberry [27]: $K(N_j) = \sqrt{N_j}$ where N_j is the number of labeled observations in the extended object training data set for object class O_j as presented in Table 2.1.

feature vectors, we will create a scatter plot consisting of the average error rates produced by the single classifier on the two error estimation methods. We will use the scatter plots to determine the feature vector in each dimension that minimize both the average error rates and absolute difference between the average error rates for the single classifier on the two error estimation methods. The combination of classifiers and error estimation methods considered in this analysis are: (KNN classifier (with holdout method), KNN classifier (with leave-one-out method)) and (Parzen classifier (with holdout method), Parzen classifier (with leave-one-out method)) and (Bayesian classifier (with holdout method), Bayesian classifier (with resubstitution method)). Table 4.15 (a–c) presents the minimum average error rates with the smallest absolute difference in the error rates on the test data set for each combination across each dimension. After identifying the minimum average error rates in each dimension and combination, we can compare the results and select

142   4 Thermal Feature Selection

Table 4.16 Compact object set of most favorable feature vectors for combinations of a classifier and error estimation methods.

FEATURE VECTOR	CLASSIFIER	ERROR ESTIMATION METHOD	AVERAGE ERROR RATES (%)		K VALUE	h VALUE
6 7 9 14	KNN	(Holdout, Leave-one-out)	7.2917	7.8616	9	
6 7 12 14	Bayesian	(Holdout, Resubstitution)	7.2917	7.3899	*	
1 6 7 12 15	KNN	(Holdout, Leave-one-out)	7.2917	7.0755	7	
6 7 9 12 14	Bayesian	(Holdout, Resubstitution)	7.2917	7.5472	*	
4 5 6 7 12 15	KNN	(Holdout, Leave-one-out)	7.2115	7.7044	5	
1 6 7 9 10 14	Bayesian	(Holdout, Resubstitution)	6.2500	6.7610	*	
1 4 6 7 9 10 14	KNN	(Holdout, Leave-one-out)	7.2917	7.3899	6	
1 6 7 8 9 10 14	Parzen	(Holdout, Leave-one-out)	7.2115	7.5472		0.0653
1 2 6 7 10 12 14	Bayesian	(Holdout, Resubstitution)	7.2917	7.5472	*	
1 3 4 5 6 12 14 15	KNN	(Holdout, Leave-one-out)	7.2917	7.5472	7	
1 6 7 8 9 10 11 14	Parzen	(Holdout, Leave-one-out)	7.2115	7.5472		0.0650
3 4 6 7 9 10 14 15	Bayesian	(Holdout, Resubstitution)	6.2500	6.7610	*	
1 3 4 5 6 7 8 10 14	KNN	(Holdout, Leave-one-out)	7.2917	7.3899	6	
1 3 6 7 9 10 13 14 15	Bayesian	(Holdout, Resubstitution)	6.2500	6.9182	*	
1 3 4 5 6 7 9 10 11 14	KNN	(Holdout, Leave-one-out)	7.2917	7.0755	6	
1 2 3 6 7 9 10 13 14 15	Bayesian	(Holdout, Resubstitution)	6.2500	6.9182	*	
1 3 4 5 6 7 8 9 11 14 15	KNN	(Holdout, Leave-one-out)	7.2917	7.7044	6	
1 2 3 4 6 7 8 10 12 13 14	Bayesian	(Holdout, Resubstitution)	7.2917	7.5472	*	
1 2 3 4 5 6 7 9 10 11 14 15	KNN	(Holdout, Leave-one-out)	7.2917	7.5472	5	
1 3 4 5 6 7 8 9 10 11 12 14 15	KNN	(Holdout, Leave-one-out)	7.2917	7.7044	3	

* Bayesian classifier will use the functional form of K in terms of the training data size that was presented by Loftsgaarden and Quesenberry [27]: $K(N_j) = \sqrt{N_j}$ where N_j is the number of labeled observations in the extended object training data set for object class Oj as presented in Table 2.1.

Table 4.17 Compact object set of most favorable feature vectors (combined feature vectors from Tables 4.14 and 4.16).

(a)

FEATURE VECTOR	CLASSIFIER	ERROR ESTIMATION METHOD	AVERAGE ERROR RATES (%)		K VALUE	h VALUE
6 7 9 14	KNN	(Holdout, Leave-one-out)	7.2917	7.8616	9	
6 7 12 14	Bayesian	(Holdout, Resubstitution)	7.2917	7.3899	*	
1 6 7 12 15	KNN	(Holdout, Leave-one-out)	7.2917	7.0755	7	
6 7 9 12 14	Bayesian	(Holdout, Resubstitution)	7.2917	7.5472	*	
1 3 6 7 12 14	Bayesian	Resubstitution	6.4465		*	
4 6 8 9 11 13	Bayesian	Holdout	2.0032		*	
3 6 8 9 11 14	Parzen	Holdout	3.0449			0.060989
4 5 6 7 12 15	KNN	(Holdout, Leave-one-out)	7.2115	7.7044	5	
1 6 7 9 10 14	Bayesian	(Holdout, Resubstitution)	6.2500	6.7610	*	
1 2 6 7 9 10 14	Bayesian	Resubstitution	6.4465		*	
1 6 7 11 12 13 15	Bayesian	Holdout	2.0032		*	
3 6 8 9 10 11 14	KNN	Holdout	4.0064		15	
1 4 6 7 9 10 14	KNN	(Holdout, Leave-one-out)	7.2917	7.3899	6	
1 6 7 8 9 10 14	Parzen	(Holdout, Leave-one-out)	7.2115	7.5472		0.0653
1 2 6 7 10 12 14	Bayesian	(Holdout, Resubstitution)	7.2917	7.5472	*	
1 2 4 6 7 9 13 14	Bayesian	Resubstitution	6.4465		*	
1 6 7 9 11 12 13 14	Bayesian	Holdout	2.0032		*	
1 5 6 7 9 10 12 14	KNN	Leave-one-out	5.8176		5	
1 3 4 5 6 12 14 15	KNN	(Holdout, Leave-one-out)	7.2917	7.5472	7	
1 6 7 8 9 10 11 14	Parzen	(Holdout, Leave-one-out)	7.2115	7.5472		0.0650
3 4 6 7 9 10 14 15	Bayesian	(Holdout, Resubstitution)	6.2500	6.7610	*	
1 2 4 6 7 9 10 13 14	Bayesian	Resubstitution	6.4465		*	
1 6 7 8 10 11 12 13 14	Bayesian	Holdout	2.0032		*	
1 5 6 7 9 10 11 12 14	KNN	Leave-one-out	5.8176		5	
1 3 4 5 6 7 8 10 14	KNN	(Holdout, Leave-one-out)	7.2917	7.3899	6	
1 3 6 7 9 10 13 14 15	Bayesian	(Holdout, Resubstitution)	6.2500	6.9182	*	

(b)

FEATURE VECTOR												CLASSIFIER	ERROR ESTIMATION METHOD	AVERAGE ERROR RATES (%)		K VALUE	h VALUE
1	6	7	9	10	11	12	13	14	15			Bayesian	Holdout	2.0032		*	
1	3	4	5	6	7	9	10	11	14			KNN	(Holdout, Leave-one-out)	7.2917	7.0755	6	
1	2	3	6	7	9	10	13	14	15			Bayesian	(Holdout, Resubstitution)	6.2500	6.9182	*	
1	2	4	6	7	8	10	11	13	14	15		Bayesian	Holdout	2.0032		*	
1	3	4	5	6	7	8	9	11	14	15		KNN	(Holdout, Leave-one-out)	7.2917	7.7044	6	
1	2	3	4	6	7	8	10	12	13	14		Bayesian	(Holdout, Resubstitution)	7.2917	7.5472	*	
1	2	3	4	5	6	7	9	10	11	14	15	KNN	(Holdout, Leave-one-out)	7.2917	7.5472	5	
1	3	4	5	6	7	8	9	10	11	12	14	15	KNN	(Holdout, Leave-one-out)	7.2917	7.7044	3

* Bayesian classifier will use the functional form of K in terms of the training data size that was presented by Loftsgaarden and Quesenberry [27]: $K(N_j) = \sqrt{N_j}$ where N_j is the number of labeled observations in the extended object training data set for object class Oj as presented in Table 2.1.

the most favorable feature vectors associated with the lowest error rates. Table 4.16 displays a set of most favorable feature vectors for the combinations of a classifier and error estimation methods.

Tables 4.17 (a–b) combine the results from Tables 4.14 and 4.16 to present our set of most favorable feature vectors for the compact objects. As in the case with the extended objects, none of the most favorable feature vectors for the compact objects consist of only features from a single feature category – meteorological, micro, or macro. Also, we are again considering a set of most favorable feature vectors that will allow us to design a classification model that is able to generalize to other test (or blind) data sets, rather than choosing a single feature vector that results in a model with too little flexibility.

4.6 Sensitivity Analysis

In the previous section, we identified sets of most favorable feature vectors for our extended and compact objects. We will now analyze the effects of variations in the camera's viewing angle, window size of the thermal scene, and rotational orientation of the target on the feature values and classification performance of a classifier involving selected feature vectors from our most favorable sets. Before we begin we will specify our rules of engagement for this analysis. Since one of our objectives is to study the behavior of the features with the noted variations, we will only make within class inferences and will not present conclusions from between class comparisons. Furthermore, we will explore the effects of these variations on classification performance within each class. However, we will not attempt to make inferences on the misclassifications until Chap. 5. All images used to generate the features were captured at a distance of 2.4 meters between the thermal camera and target. The images were processed as discussed in Chap. 2 and the features were generated as presented in Chap. 3.

4.6.1 Viewing Angle Variations

The sensitivity analysis for the variations in the camera's viewing angle will be performed using the extended objects and the Bayesian classifier with the 9-dimensional extended object feature vector < 2, 3, 5, 6, 7, 8, 9, 11, 12 >. The features associated with the numerical labels in the feature vector are presented in Table 4.2. This classification model displayed an error rate of approximately 2.95% with the holdout error estimation method on the extended objects in Sect. 4.5.5. The extended objects used in the analysis consist of a brick wall, hedges, picket fence, and wood wall. The thermal images were captured on 10 and 11 February 2007 between 1300 and 1700 hrs on each day with meteorological conditions involving clear skies and temperatures ranging from approximately $42.2°F$ to $49.8°F$. This temperature range influenced our choice for the most favorable feature vector used in this analysis. Consequently, we did not choose a feature vector that included the ambient temperature feature since minimal data is available in the training set for this range of temperatures as shown in Fig. 2.15. The thermal images of these extended objects were captured at seven different viewing angles: $-60°$ from normal incidence, $-45°$ from normal incidence, $-30°$ from normal incidence, normal incidence ($0°$), $30°$ from normal incidence, $45°$ from normal incidence, and $60°$ from normal incidence. The visible image and thermal images for each viewing angle is presented for each object in Fig. 4.14.

Table 4.18 presents the feature values and assigned classes as well as the posterior probabilities of the Bayesian classifier for each extended object with variations in the camera's viewing angle. The object surface radiance (Lo) feature values show strong variations for both the picket fence and wood wall. The background irradiance (Lb) values display variations as expected for each object class since thermal energy from different background sources is being reflected diffusely from the aluminum foil as the camera varies its viewing angle. In the context of this research, we have defined background as the region either in front or to the side of the target consisting of thermal sources that emit thermal energy onto the target's surface. The source emitting this thermal energy may or may not be in the camera's field of view. On the other hand, we have defined foreground as the region in the scene consisting of objects behind the target of interest and within the thermal camera's field of view. In Sect. 3.5.2, we noted that for nonmetallic materials such as wood and vegetation, the emissivity remains rather constant across variations in the viewing angle up to about $50°$ from normal incidence [33]. This statement appears to hold true for the picket fence and somewhat true for the other three object classes. However, we would not expect the emissivity feature values (or any other feature values) for our object classes to be well behaved as they would in a controlled laboratory since our objects' feature values depend on a dynamical outdoor environment. The reference emitter's

4.6 Sensitivity Analysis   145

Table 4.18 Variations in the camera's viewing angle effect on feature values and classification performance of a Bayesian classifier for each extended object in the left column. The object class assigned by the classifier as well as the posterior probabilities for each object class is presented in the columns on the right.

OBJECT	ANGLE FROM INCIDENCE (Degrees)	FEATURE VALUES										Assigned Class	POSTERIOR PROBABILITIES			
		T1	Lo	Lb	Lor	Lob	Eo	Mo1	Mob1	So1	Lr		Brick Wall	Hedges	Picket Fence	Wood Wall
Brick Wall	-60	0.0433	94.9617	92.9803	0.9193	1.0213	0.1864	94.9844	1.0216	8.1730E-05	103.2929	Brick Wall	0.7756	0.2242	0.0001	3.7972E-05
	-45	0.0433	94.7209	94.1202	0.8907	1.0064	0.0477	94.8433	1.0077	7.9345E-05	106.3387	Hedges	0.2486	0.7314	0.0200	1.9581E-05
	-30	0.0433	94.5467	94.5482	0.8950	1.0000	-0.0001	94.5213	0.9997	8.0769E-05	105.6375	Brick Wall	0.5810	0.3396	0.0793	6.4467E-05
	0	0.0433	94.7672	95.9912	0.8898	0.9872	-0.1129	94.7900	0.9875	6.7887E-05	106.5030	Brick Wall	0.6203	0.3532	0.0264	6.3010E-05
	30	0.0433	94.2989	95.0153	0.8876	0.9925	-0.0619	94.2612	0.9921	7.6643E-05	106.2397	Brick Wall	0.7249	0.2020	0.0730	1.7991E-04
	45	0.0433	94.7160	95.3002	0.8886	0.9939	-0.0502	94.7553	0.9943	9.5116E-05	106.5842	Brick Wall	0.4744	0.4587	0.0668	4.0784E-05
	60	0.0433	94.8038	92.7395	0.8848	1.0223	0.1390	94.8909	1.0232	6.5737E-05	107.1417	Hedges	0.1465	0.8534	0.0001	2.0882E-06
Hedges	-60	0.0133	95.2910	100.5700	0.8144	0.9475	-0.3116	94.8174	0.9428	6.5982E-04	117.0042	Hedges	0.0216	0.9762	0.0020	0.0002
	-45	0.0133	95.3998	96.0246	0.7517	0.9935	-0.0196	95.0911	0.9903	7.1898E-04	126.9160	Hedges	0.0011	0.7670	0.2318	1.7443E-05
	-30	0.0133	95.6480	94.9257	0.7218	1.0076	0.0186	95.1725	1.0026	7.2689E-04	132.5179	Hedges	0.0007	0.6015	0.3978	6.3578E-06
	0	0.0133	95.7860	95.8691	0.7297	0.9991	-0.0023	95.2173	0.9932	6.3364E-04	131.2684	Hedges	0.0025	0.7604	0.2371	1.0913E-05
	30	0.0133	95.1127	97.4163	0.7185	0.9764	-0.0639	95.1133	0.9764	4.4896E-04	132.3797	Hedges	0.0049	0.9895	0.0056	1.9066E-06
	45	0.0133	95.3904	97.2083	0.7879	0.9813	-0.0739	95.1845	0.9792	2.0106E-04	121.0743	Hedges	0.0521	0.9389	0.0090	5.4414E-06
	60	0.0133	94.9263	94.7073	0.7130	1.0023	0.0055	95.0170	1.0033	1.0357E-04	133.1368	Hedges	0.0004	0.9990	0.0006	1.4435E-08
Picket Fence	-60	0.0233	124.7094	90.2790	0.8523	1.3814	0.5958	94.1866	1.0433	9.8058E-03	146.3294	Picket Fence	2.1317E-08	2.2893E-07	0.8101	0.1899
	-45	0.0233	129.0326	90.2098	0.8338	1.4304	0.5835	94.1903	1.0441	1.9036E-02	154.7484	Picket Fence	8.2947E-09	7.7560E-07	0.5769	0.4231
	-30	0.0233	124.1194	93.4587	0.8171	1.3281	0.5089	94.8792	1.0152	9.6040E-03	151.8996	Picket Fence	9.0397E-09	9.6969E-08	0.9199	0.0801
	0	0.0233	125.5684	95.2828	0.8673	1.3179	0.5935	94.4753	0.9915	1.4192E-02	144.7846	Picket Fence	7.2086E-09	3.4656E-07	0.7309	0.2691
	30	0.0233	120.1833	93.0409	0.8333	1.2917	0.5144	92.2863	0.9919	1.5219E-02	144.2259	Picket Fence	6.6776E-09	8.2017E-07	0.6884	0.3116
	45	0.0233	127.0912	92.7001	0.8525	1.3710	0.5916	93.5688	1.0094	1.5773E-02	149.0874	Picket Fence	7.2573E-09	4.2859E-07	0.7070	0.2930
	60	0.0233	128.7721	90.1232	0.8536	1.4288	0.6172	93.5213	1.0377	1.2437E-02	150.8625	Picket Fence	1.8253E-08	3.2385E-07	0.7682	0.2318
Wood Wall	-60	-0.0200	103.0582	92.3384	0.8384	1.1161	0.3400	94.5762	1.0242	2.0918E-03	122.9228	Picket Fence	2.3124E-07	7.5991E-07	0.9942	0.0058
	-45	-0.0200	97.1941	97.1622	0.7699	1.0003	0.0011	95.0557	0.9783	2.4461E-03	126.2458	Picket Fence	1.2691E-06	0.0010	0.9978	0.0012
	-30	-0.0200	99.1662	96.1854	0.7882	1.0310	0.0976	94.7222	0.9848	3.0110E-03	125.8145	Picket Fence	6.7161E-08	2.0578E-05	0.9989	0.0011
	0	-0.0200	96.9300	96.5391	0.8133	1.0040	0.0167	95.0429	0.9845	3.4981E-03	119.1762	Picket Fence	6.0568E-08	8.3124E-05	0.9928	0.0072
	30	-0.0200	97.8098	95.1851	0.8065	1.0276	0.0976	94.4559	0.9923	3.4576E-03	121.2698	Picket Fence	2.0069E-08	1.9857E-05	0.9972	0.0028
	45	-0.0200	95.9054	95.3208	0.8026	1.0061	0.0235	94.9616	0.9962	3.7237E-03	119.4957	Picket Fence	2.1684E-08	4.1212E-05	0.9961	0.0039
	60	-0.0200	96.3324	90.3376	0.8043	1.0664	0.1975	94.9803	1.0514	3.0500E-03	119.7736	Picket Fence	2.6450E-06	0.0059	0.9214	0.0727

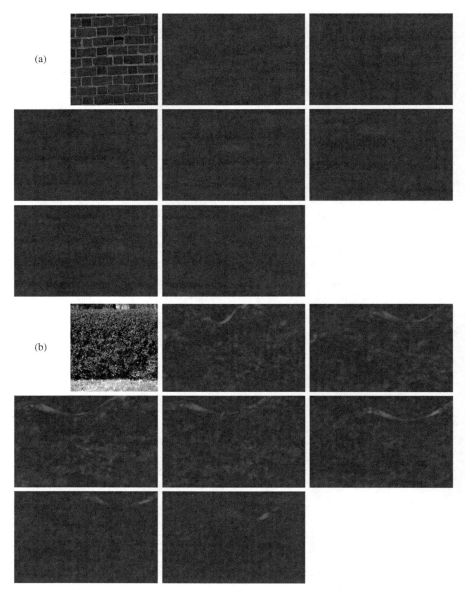

Fig. 4.14 Visible images and thermal images for each viewing angle of extended objects used in sensitivity analysis for the variations in the camera's viewing angle. The viewing angles of the thermal images are arranged from left to right as −60° from normal incidence, −45° from normal incidence, −30° from normal incidence, normal incidence, 30° from normal incidence, 45° from normal incidence, and 60° from normal incidence. (a) brick wall, (b) hedges, (c) picket fence, (d) wood wall.

4.6 Sensitivity Analysis 147

(c)

(d)

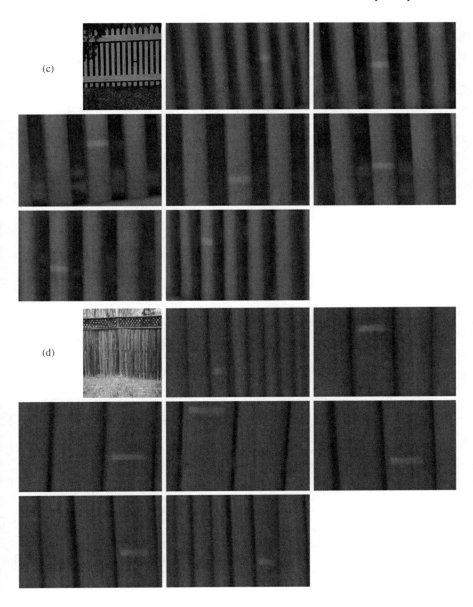

radiance (Lr) displays a large deviation in its values for the hedges due to the electrical tape being attached to an irregular surface. The variation in the camera's viewing angle has a strong effect on the posterior probabilities for the brick wall as seen by the variation in the resulting probabilities of a brick wall and hedges. We see a moderate effect on the posterior probabilities for the picket fence as seen by the variations in the resulting probabilities of a picket fence and wood wall. The variation in the camera's viewing angle appears to have a minor effect on the posterior probabilities for the hedges and wood wall object classes. As mentioned earlier, we will discuss reasons for misclassifications of objects by a classification model in Chap. 5. In general, we conclude that variations in the viewing angle of a thermal camera will have a moderate effect on the values of features and performance of a classification model.

4.6.2 Window Size Variations

The sensitivity analysis for the variations in the window size of the thermal scene will be performed using the extended objects and the Bayesian classifier with the 9-dimensional extended object feature vector < 2, 3, 5, 9, 10, 11, 13, 15, 17 >. The features associated with the numerical labels in the feature vector are presented in Table 4.2. This classification model displayed estimated error rates of 18.06% with the holdout error estimation method and 17.91% with the resubstitution error estimation method on the extended objects in Sect. 4.5.5. The extended objects used in the analysis consist of a brick wall, hedges, picket fence, and wood wall. The thermal images of the extended objects were captured on 10 and 11 February 2007 between 0930 and 1700 hrs on each day with meteorological conditions involving clear skies and temperatures ranging from approximately $42.2°F$ to $49.8°F$. Consequently, we did not choose a most favorable feature vector with a temperature feature since minimal data is available in the training set for this range of temperatures as shown in Fig. 2.15. Furthermore, we chose a feature vector having a majority of macro features since we will vary the window size of the entire scene of the target with the micro features generated from each target's surface remaining constant. For each extended object used in the analysis, the window size of the thermal scene containing the target decreases in increments to produce 100 thermal images that are each proportional to the original segment. As previously mentioned, the values for the micro feature Lo as well as Lb and the meteorological feature $T1$ will remain constant for each object's images. However, the values for the macro features $Mo1$, $Mor1$, $Mob1$, $En1$, $Cr2$, and $Ho2$ will be computed for each window size. The visible image and thermal images to include the first (largest) and 100th (smallest) window segment for each of these objects is displayed in Fig. 4.15.

For the brick wall, the constant feature values were $T1 = 0.05$, $Lo = 95.0405$, and $Lb = 94.4728$. Figure 4.16 (a) presents the posterior probabilities for the brick wall feature vectors being a brick wall, hedges, picket fence, and wood wall. The posterior probabilities of the brick wall and wood wall display minimal variations. However,

4.6 Sensitivity Analysis

Fig. 4.15 Visible images and thermal images for extended objects used in sensitivity analysis for the variations in the window size of the thermal scene. The first (largest) and 100th (smallest) window segments out of the 100 window sizes are enclosed by the solid red borders. (a) brick wall, (b) hedges, (c) picket fence, (d) wood wall.

there is minimal variation in the classifier's posterior probabilities of the hedges and picket fence until about the 80th window size index. For the macro feature values in Fig. 4.16 (b), the largest variations occur at around the 80th window size index. However, the features $En1$ and $Cr2$ display a gradual change in values up to the 80th window size index.

For the hedges, the constant feature values were $T1 = 0.0567$, $Lo = 94.0763$, and $Lb = 97.4769$. Figure 4.17 (a) presents the posterior probabilities for the hedges feature vectors being a brick wall, hedges, picket fence, and wood wall. The posterior probabilities of the brick wall and wood wall display minimal variations. However, there is minimal variation in the classifier's posterior probabilities of the hedges and picket fence until about the 90th window size index. For the macro feature values in Fig. 4.17 (b), the largest variations occur around the 90th window size index. The features $Mo1$, $Mor1$, and $Mob1$ display a gradual change in values up to about the 90th window size index.

For the picket fence, the constant values were $T1 = 0.0233$, $Lo = 123.221$, and $Lb = 94.996$. Figure 4.18 (a) displays the posterior probabilities for the picket fence feature vectors being a brick wall, hedges, picket fence, and wood wall. The posterior probabilities of the brick wall and hedges display minimal variations. However, there is minimal variation in the classifier's posterior probabilities of the picket fence and wood wall until about the 80th window size index. For the macro feature values in Fig. 4.18 (b), the largest variations occur around the 90th window size index. Additionally, all the macro features for the picket fence display a gradual change in values up to about the 90th window size index.

For the wood wall, the constant values were $T1 = -0.02$, $Lo = 96.8051$, and $Lb = 97.566$. Figure 4.19 (a) displays the posterior probabilities for the wood wall feature vectors being a brick wall, hedges, picket fence, and wood wall. The posterior probabilities of the brick wall and hedges display minimal variations. However, there is minimal variation in the classifier's posterior probabilities of the picket fence and wood wall until about the 90th window size index. For the macro feature values in Fig. 4.19 (b), the largest variations occur around the 90th window size index. With the exception of $Ho2$, all the macro features for the wood wall display a gradual change in values up to about the 90th window size index. The $Ho2$ macro feature remains approximately constant until the 90th window size index.

In general, we conclude that the variations in the window size of the thermal scene of a target will have a moderate effect on the values of features and minor effect on the posterior probabilities of a classification model. As we see, these variations in the window size of the thermal scene of a target could act as a *dynamical window* technique that affords an autonomous robot the ability to collect multiple degrees of information regarding a target's surface to arrive at a more confident decision for a class assignment. We use this dynamical window technique in our novel classification model that we will present in Chap. 5.

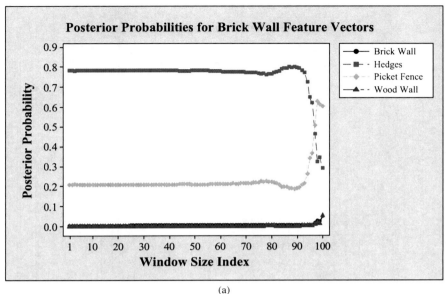

(a)

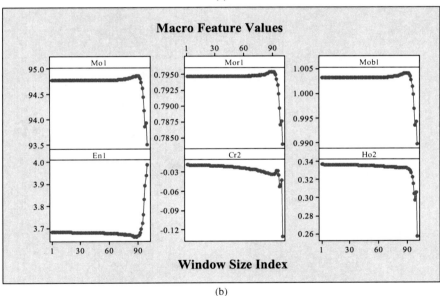

(b)

Fig. 4.16 Brick wall sensitivity analysis for the variations in the window size of the thermal scene. (a) Posterior probabilities for the brick wall feature vectors and (b) macro feature values with variations in window size indexed from 1 (largest window) to 100 (smallest window).

152 4 Thermal Feature Selection

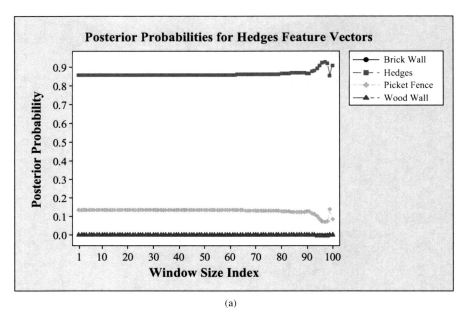

(a)

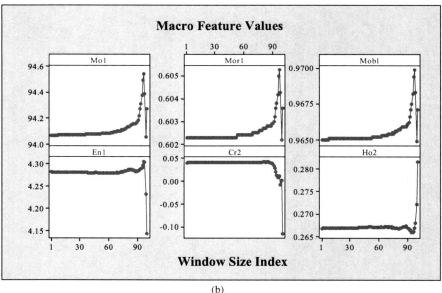

(b)

Fig. 4.17 Hedges sensitivity analysis for the variations in the window size of the thermal scene. (a) Posterior probabilities for the hedges feature vectors and (b) macro feature values with variations in window size indexed from 1 (largest window) to 100 (smallest window).

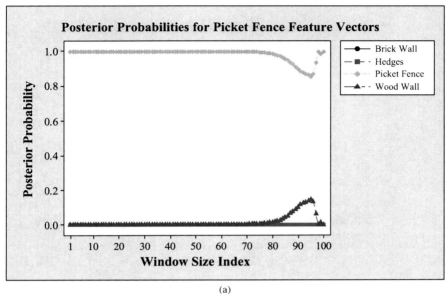

(a)

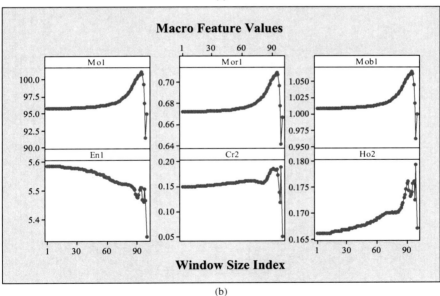

(b)

Fig. 4.18 Picket fence sensitivity analysis for the variations in the window size of the thermal scene. (a) Posterior probabilities for the picket fence feature vectors and (b) macro feature values with variations in window size indexed from 1 (largest window) to 100 (smallest window).

154   4 Thermal Feature Selection

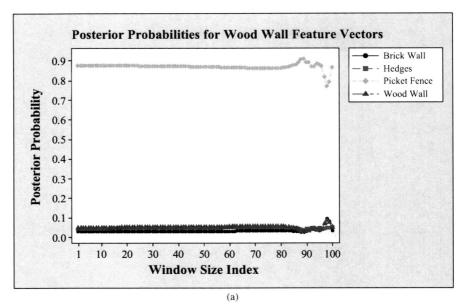

(a)

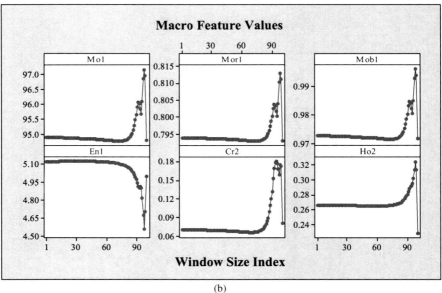

(b)

Fig. 4.19 Wood wall sensitivity analysis for the variations in the window size of the thermal scene. (a) Posterior probabilities for the wood wall feature vectors and (b) macro feature values with variations in window size indexed from 1 (largest window) to 100 (smallest window).

4.6.3 Rotational Variations

The sensitivity analysis for the variations in the rotational orientation will be performed using a compact object and the Bayesian classifier with the 9-dimensional compact object feature vector < 1, 6, 7, 8, 10, 11, 12, 13, 14 >. The features associated with the numerical labels in the feature vector are presented in Table 4.10. This classification model displayed an estimated error rate of 2.00% with the holdout error estimation method on the compact objects in Sect. 4.5.6. The compact object used in this analysis is a pine tree log with the thermal image captured on 9 October 2007 at 1317 hrs with meteorological conditions involving clear skies and an ambient temperature of approximately $98.2°F$. The thermal image of the pine tree log was rotated to produce images with five different orientations: 0°, 45°, 90°, 135°, and 180°. The visible image and thermal images for each orientation along with the segmented portion of the pine tree log used in the analysis is presented in Fig. 4.20.

Table 4.19 presents the feature values and assigned classes as well as the posterior probabilities of the Bayesian classifier for each orientation of the pine tree log. The feature values for *Eo* display a large deviation at the diagonal angles (45° and 135°) from the feature values found for 0°, 90°, and 180°. Additionally, *Co2* shows a large variation in values at the angles 135° and 180° from those feature values of the other three angles. However, the other feature values display minimal variations with the rotational angles. As we see, the variation in the orientation of the pine tree log has a minimal effect on the posterior probabilities. Consequently, we can conclude that variations in the rotational orientation of an object have a minor effect on the values of the features and posterior probabilities of a classification model. Therefore, the performance of the classification model is rotational invariant.

Table 4.19 Effect of variations in the rotational orientation on feature values and classification performance of a Bayesian classifier of a pine tree log. The object class assigned by the classifier as well as the posterior probabilities for each rotation angle is presented in the columns on the right.

OBJECT	ROTATION ANGLE (Degrees)	FEATURE VALUES									Assigned Class	POSTERIOR PROBABILITIES	
		Ta	Lor	Lob	Eo	En1	Co2	Cr2	Er2	Ho2		Steel Pole	Tree
Pine Tree Log	0	98.2	1.0056	1.0407	1.1295	3.8779	15.2746	0.5182	0.0102	0.3866	Tree	0.0002	0.9998
	45	98.2	1.0273	1.0314	7.6905	3.9113	14.0655	0.6843	0.0124	0.3858	Tree	0.0202	0.9798
	90	98.2	1.0031	1.0358	1.0636	3.8529	14.0995	0.6134	0.0111	0.3982	Tree	0.0004	0.9996
	135	98.2	1.0359	1.0494	3.6732	4.4568	21.5122	0.7750	0.0067	0.3477	Tree	0.0002	0.9998
	180	98.2	1.0176	1.0462	1.5940	4.3081	20.3509	0.6557	0.0066	0.3576	Tree	0.0001	0.9999

156   4 Thermal Feature Selection

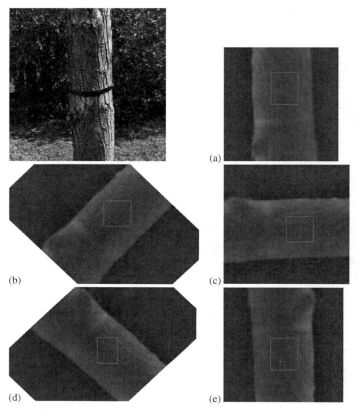

Fig. 4.20 Visible image and thermal images for the pine tree log used in the sensitivity analysis for the variations in the rotational orientation. (a) 0°, (b) 45°, (c) 90°, (d) 135°, (e) 180°. The portion of the pine tree log segmented for the analysis is enclosed by the solid rectangular borders in each thermal image.

4.7 Summary

In this chapter, we evaluated the performance of various classification models to identify the most favorable feature vectors for our extended and compact objects. We first introduced our approach of statistical pattern classification where learning involves labeled classes of data (supervised classification), assumes no formal structure regarding the density of the data in the classes (nonparametric density estimation), and makes direct use of posterior probabilities when making decisions regarding class assignments (probabilistic approach). After presenting a preliminary feature analysis to assess the quality of our training data and eliminate redundant features, classification models were formed with feature vectors from the extended and compact object classes and Bayesian, KNN, and Parzen classifiers. The error rates for each classification model were computed using exhaustive search feature selection method on a high performance computing system. For each classification model, an average error rate was computed on test set data using the resubstitution, holdout, and leave-one-out error estimation methods. The resulting average error rates were compared to determine the feature vectors that present the lowest error rates. These feature vectors were considered as our most favorable feature vectors and consist of relevant information to allow us to classify unknown non-heat generating objects with minimal error. We saw that there is no single "optimal" feature vector but a set of "most favorable" feature vectors associated with various classifiers for both the extend and compact object classes. Moreover, we showed that the most favorable feature vectors are those that contain contributions from all the feature types – meteorological, micro, and macro.

We performed a sensitivity analysis to explore the effects of variations in the camera's viewing angle, window size of the thermal scene, and rotational orientation of the target on the feature values and classification performance of a classifier involving selected feature vectors from our most favorable sets. In general, we conclude that variations in the viewing angle of a thermal camera will have a moderate effect on the values of features and performance of a classification model. The variations in the window size of the thermal scene of a target have a moderate effect on the values of features and minor effect on the posterior probabilities of a classification model. Additionally, we concluded that variations in the rotational orientation of an object have a minor effect on the values of the features and posterior probabilities of a classification model.

During the sensitivity analysis in Sect. 4.6, we noted that the variations in the window size of the thermal scene of a target could act as a *dynamical window* technique that affords an autonomous robot the ability to collect multiple degrees of information regarding a target's thermal scene to arrive at a more confident decision for a class assignment. We use this dynamical window technique in our novel classification model that we will present in Chap. 5. Furthermore, we saw that some patterns from specific object classes were consistently misclassified while other patterns were assigned to the correct class. In Chap. 5, we will identify conditions that result in blind patterns from specific object classes being misclassified. Additionally, we will observe that certain classification models perform exceptionally well on patterns from specific object classes. These classification models act as *experts* in making classification decisions for patterns from these respective object classes. It turns out that we can form a *committee of experts* for classifying patterns from a specific object class. By combining each committee of experts into one classification model, we are able to exploit the expertise of each committee and complement the overall performance of the classification model. This novel concept is the baseline for our *Adaptive Bayesian Classification Model* presented next in Chap. 5.

References

[1] Duda RO, Hart PE et al (2001) Pattern Classification. 2nd edn. Wiley, New York
[2] Jain AK, Duin RPW et al (2000) Statistical Pattern Recognition: A Review. IEEE Transactions on Pattern Analysis and Machine Intelligence 22(1):4–37
[3] Polikar R (2006) Pattern Recognition. In: Akay M (ed) Wiley Encyclopedia of Biomedical Engineering. Wiley-Interscience; John Wiley, Hoboken, N.J.; Chichester
[4] Bishop CM (1995) Neural Networks for Pattern Recognition. Clarendon Press; Oxford University Press, Oxford; New York
[5] Theodoridis S, Koutroumbas K (2006) Pattern Recognition. 3rd edn. Academic Press, San Diego, CA
[6] van der Heijden F, Duin RPW et al (2004) Classification, Parameter Estimation, and State Estimation : An Engineering Approach using MATLAB. Wiley, Chichester, West Sussex, Eng. ; Hoboken, NJ
[7] Webb AR (2002) Statistical Pattern Recognition. 2nd edn. Wiley, West Sussex, England ; New Jersey
[8] Patrick EA (1972) Fundamentals of Pattern Recognition. Prentice-Hall, Englewood Cliffs, N.J.
[9] Duda RO, Hart PE (1973) Pattern Classification and Scene Analysis. Wiley, New York
[10] Hand DJ (1981) Discrimination and Classification. Wiley, Chichester Eng. ; New York
[11] Devijver PA, Kittler J (1982) Pattern Recognition : A Statistical Approach. Prentice/Hall International, Englewood Cliffs, N.J.
[12] Fukunaga K (1990) Introduction to Statistical Pattern Recognition. 2nd edn. Academic Press, Boston
[13] Devroye L, Györfi L et al (1996) A Probabilistic Theory of Pattern Recognition. Springer, New York
[14] Dasarathy BV (1991) Nearest Neighbor (NN) Norms : Nn Pattern Classification Techniques. IEEE Computer Society Press; IEEE Computer Society Press Tutorial, Los Alamitos, Calif.; Washington
[15] Duin RPW, Juszczak P et al (2004) PRTools4, A Matlab Toolbox for Pattern Recognition. Delft University of Technology, The Netherlands
[16] Milligan GW, Cooper MC (1988) A Study of Standardization of Variables in Cluster Analysis. Journal of Classification 5:181–204
[17] Burnham KP, Anderson DR (2002) Model Selection and Multimodel Inference : A Practical Information-Theoretic Approach. 2nd edn. Springer, New York
[18] Holst GC (2000) Common Sense Approach to Thermal Imaging. JCD Pub.; co-published by SPIE Optical Engineering Press, Winter Park, Fla.; Bellingham, Wash.
[19] Buluswar SD, Draper BA (1998) Color Machine Vision for Autonomous Vehicles. Engineering Applications of Artificial Intelligence(11):245–256
[20] Buluswar SD, Draper BA (2002) Color Models for Outdoor Machine Vision. Computer Vision and Image Understanding 85(2):71–99
[21] Manduchi R, Castano A et al (2005) Obstacle Detection and Terrain Classification for Autonomous Off-Road Navigation. Autonomous Robots 18(1):81–102
[22] Fisher RA (1922) On the Mathematical Foundations of Theoretical Statistics. Philosophical Transactions of the Royal Society of London.Series A, Containing Papers of a Mathematical Or Physical Character 222:309–368
[23] Fix, E., and Hodges, J. L. (1951). *Discriminatory Analysis: Nonparametric Discrimination: Consistency Properties,* USAF School of Aviation Medicine, Randolph Field, Texas
[24] Fix, E., and Hodges, J. L. (1952). *Discriminatory Analysis: Nonparametric Discrimination: Small Sample Performance,* USAF School of Aviation Medicine, Randolph Field, Texas
[25] Cover TM, Hart PE (1967) Nearest Neighbor Pattern Classification. IEEE Transactions on Information Theory IT-13(1):21–27

[26] Parzen E (1962) On Estimation of a Probability Density Function and Mode. The Annals of Mathematical Statistics 33(3):1065–1076
[27] Loftsgaarden DO, Quesenberry CP (1965) A Nonparametric Estimate of a Multivariate Density Function. The Annals of Mathematical Statistics 36(3):1049–1051
[28] Fukunaga K, Hostetler L (1973) Optimization of k Nearest Neighbor Density Estimates. IEEE Transactions on Information Theory 19(3):320–326
[29] Enas GG, Choi SC (1986) Choice of the Smoothing Parameter and Efficiency of k-Nearest Neighbor Classification. Computers & Mathematics with Applications, 12A(2):235–244
[30] Ghosh AK (2006) On Optimum Choice of k in Nearest Neighbor Classification. Computational Statistics & Data Analysis 50(11):3113–3123
[31] Hyvarinen A, Karhunen J et al (2001) Independent Component Analysis. J. Wiley, New York
[32] Vicente MA, Hoyer PO et al (2007) Equivalence of some Common Linear Feature Extraction Techniques for Appearance-Based Object Recognition Tasks. IEEE Transactions on Pattern Analysis and Machine Intelligence 29(5):896–900
[33] Maldague XPV (2001) Theory and Practice of Infrared Technology for Nondestructive Testing. Wiley, New York

5 Adaptive Bayesian Classification Model

Abstract An Adaptive Bayesian Classification Model is presented that outperforms the traditional KNN and Parzen classifiers when classifying non-heat generating outdoor objects in thermal scenes. The Adaptive Bayesian Classification Model is a suitable choice for any classification application involving hyperconoidal clusters. Each hyperconoidal cluster consists of an object class's patterns in an n-dimensional feature space that are characterized by their behavior about a respective first principal eigenvector. The model introduced in this chapter is designed to adapt to the behavior of these patterns from specified object classes to provide an accurate classification of unknown objects.

5.1 Introduction

In Chap. 3, we generated features from the thermal images of our non-heat generating objects. In the context of this research, we have defined non-heat generating objects as objects that are not a source for their own emission of thermal energy, and so exclude people, animals, vehicles, etc. In Chap. 4, we assessed the performance of various classification models to identify the most favorable sets of feature vectors for our extended and compact object classes. The extended objects consist of objects that extend laterally beyond the thermal camera's lateral field of view, such as brick walls, hedges, picket fences, and wood walls. The compact objects consist of objects that are completely within the thermal camera's lateral field of view, such as steel poles and trees. We will now use these most favorable feature vectors to design and implement a novel model that outperforms the traditional KNN and Parzen classifiers for our specific application. The design of the adaptive Bayesian classification model is based on the observation that the thermal patterns for each class of non-heat generating objects display a unique behavior about an eigenvector that projects through their respective hyperconoidal cluster. The behavior is characterized by the normal distances between the patterns and eigenvector for each object class. Various distance functions are derived based on these normal distances. These distance functions are incorporated into the likelihood function of the Bayesian classifiers to form an adaptive Bayesian classifier. We found that the combination of specific sets of adaptive Bayesian classifiers and most favorable feature vectors yield exceptional classification performance for a given object class. Each set of adaptive Bayesian classifier models acts as an *expert* in making classification decisions on patterns from their respective object class. It turns out that we can form a *committee of experts* for classifying patterns from a specific object class. Consequently, one committee of experts may perform exceptionally on specific unknown patterns where another classifier is deficient, and vice versa. By combining each committee of experts into one classification model, we are able to exploit the expertise of each committee and complement the overall performance of the classification model. We further increased the confidence level in our model's classification decisions by integrating the *dynamical window* technique presented in Chap. 4 that lets each committee of experts decide on class assignment by considering information collected from multiple window sizes of the thermal image of an object. Additionally, we incorporated rules into our model that must be satisfied before the bot is authorized to make a classification decision. If all the rules are satisfied, the bot is authorized to assign a class to the unknown object within its field of view and proceed with the next required action in the intelligence algorithm. On the other hand, if a rule is not satisfied, the bot must reject a class assignment and capture another thermal image of the unknown object for classification, perhaps from another viewing angle. This is the cornerstone of the Adaptive Bayesian Classification Model.

The remainder of this chapter will proceed as follows. In Sect. 5.2, we will derive the distance metrics used to describe the behavior of each object class's patterns about the eigenvector that projects through their respective hyperconoidal cluster. In Sect. 5.3, we will present our adaptive Bayesian classifiers. We will compare the classification performance of our adaptive Bayesian classifiers to the KNN and Parzen classifiers using our most favorable feature vectors on blind data sets in Sect. 5.4. We will also make inferences on blind patterns being misclassified under certain thermal conditions. In Sect. 5.5 we will present our algorithm for the Adaptive Bayesian Classification Model consisting of the committees of expert adaptive Bayesian classifiers. Section 5.6 will present an example application of our Adaptive Bayesian Classification Model. We will conclude this chapter with a summary in Sect. 5.7. The models and methods presented in this chapter were implemented using Matlab with assistance by FastICA [1] and a pattern recognition toolbox known as PRTools4 [2].

5.2 Distance Metrics for Hyperconoidal Clusters

In Chap. 4 we introduced principal component analysis (PCA) as a traditional feature extraction method for dimensionality reduction of a feature space. As

5.2 Distance Metrics for Hyperconoidal Clusters

a dimensionality reduction technique, PCA is applied globally over the patterns of all the object classes in the feature space. For a data set of size m consisting of n-dimensional feature vectors from all object classes, we showed that there exists an eigenvector that not only determines the direction of the maximal variance of the data in feature space but also best fits the patterns in a least squared sense. We refer to this eigenvector as the *first principal eigenvector*. Using PCA in a "non-traditional way" we can perform *local PCA* on each object class and compute the first principal eigenvector that provides a best fit through the respective object class's hyperconoidal cluster.

Figure 5.1 provides an example of local PCA applied to three object classes in a 3-dimensional feature space with features f1, f2, and f3. We will name these object classes red, blue, and green, corresponding to the colors in the figure. Since the patterns in the feature space illustrated in Fig. 5.1 display characteristics analogous to the patterns of our non-heat generating objects, we will generalize the following observations and equations to our extended and compact object classes. First we note that similar to the hyperconoidal clusters introduced with the 2-dimensional scatter plots in Chap. 4, the 3-dimensional hyperconoidal clusters for our non-heat generating object classes also tend to diverge from a common origin as displayed in Fig. 5.1. In Fig. 5.1, we see that the patterns about the first principal eigenvectors behave differently for each object class. Thus, two types of behavior in the patterns seem to uniquely characterize the object classes. The first type of behavior is that we see regions with dense clusters of patterns that vary in location differently for each object class. The second type of behavior is that the trend in the normal distance between each object class's patterns and their respective first principal eigenvector appears to uniquely characterize each object class.

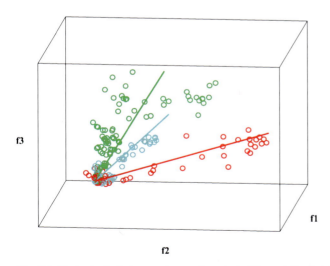

Fig. 5.1 First principal eigenvectors each projected through the hyperconoidal cluster of their respective object class in a 3-dimensional feature space.

For instance, the patterns in the blue class appear to have a more compact fit about their respective first principal eigenvector compared to the patterns in the red and green classes. The patterns in the green class appear to have larger normal distances from their respective first principal eigenvector compared to the red and blue classes. We can study these behaviors in more detail and with n-dimensional feature vectors by using the two distance metrics that we will now present.

Similar to the global PCA discussed in Chap. 4, that was applied to all the patterns in the feature space irrespective of the object class, local PCA centers the m patterns in each object class O_j, where $j = 1,...,J$, by subtracting the sample mean $\overline{f_p}$ from each feature value f_p, where $p = 1,...,n$, across each pattern $\underline{f}_{ij} = \langle f_1, f_2, ..., f_n \rangle$, where $i = 1,...,m$. This produces patterns $\underline{\tilde{f}}_{ij} = \langle \tilde{f}_1, \tilde{f}_2, ..., \tilde{f}_n \rangle$ with a mean of zero for each feature $\tilde{f}_p = f_p - \overline{f_p}$ in object class O_j. Continuing with the same procedures discussed in Chap. 4, we can compute the first principal eigenvector $\underline{e}_{1j}$ by solving the eigenvalue problem involving the covariance matrix of the matrix formed with the patterns $\underline{\tilde{f}}_{ij}$ along the columns. Repeating this process across all object classes results in hyperconoidal clusters along with their respective first principal eigenvectors that diverge from a common origin, similar to Fig. 5.1. Consequently, the first principal eigenvector $\underline{e}_{1j}$ and each pattern $\underline{\tilde{f}}_{ij}$ in object class O_j can be treated as position vectors with the common origin as the initial position.

The first distance metric that will assist us in understanding the behavior of each object class's patterns is the component (or scalar projection) of the pattern $\underline{\tilde{f}}_{ij}$ onto the first principal eigenvector $\underline{e}_{1j}$ as displayed in Fig. 5.2 and given by

$$comp_{\underline{e}_{1j}} \underline{\tilde{f}}_{ij} = \left|\underline{\tilde{f}}_{ij}\right| \cos\theta = \frac{\underline{e}_{1j} \cdot \underline{\tilde{f}}_{ij}}{\left|\underline{e}_{1j}\right|} \quad (5.1)$$

The second distance metric is the normal distance between a pattern and its respective first principal eigenvector as displayed in Fig. 5.2 and given by

$$D_{ij} = \left|\underline{\tilde{f}}_{ij}\right| |\sin\theta| \quad (5.2)$$

where $D_{ij} \geq 0$ and $\theta = \cos^{-1}\left(\frac{\underline{e}_{1j} \cdot \underline{\tilde{f}}_{ij}}{\left|\underline{e}_{1j}\right|\left|\underline{\tilde{f}}_{ij}\right|}\right)$.

5.2 Distance Metrics for Hyperconoidal Clusters

By relating $comp_{\underline{e}_{1j}} \underline{\tilde{f}}_{ij}$ to D_{ij}, we can analyze the behavior of each object class's pattern $\underline{\tilde{f}}_{ij}$ about the respective first principal eigenvector $\underline{e}_{1j}$. Figures 5.3 (a–j) present the relationships of these distance metrics involving the set of most favorable feature vectors for the extended objects presented in Table 4.9. The numerical labels in each feature vector for the extended objects are defined in Table 4.2. The training data for the extended objects displayed in Table 2.1 is used in this analysis. By comparing the plots for each object class across the given dimensions, we see that a general trend is found within each object class that varies slightly depending on the feature vector. These trends found in the relationships between the distance metrics are attributed to the values of the features for each object class within the training data set. As we saw in Chaps. 3 and 4, each object class's feature values depend on its respective material properties, the thermal camera's viewing angle, and the diurnal cycle of solar energy. These combined factors yield the trends that we see by each extended object class's patterns in Fig. 5.3. For instance, the hedges present the highest standard deviation in its feature values compared to the other extended object classes. As a result, the hedges' patterns display higher deviations across both the normal distance and scalar projection metrics in Fig. 5.3 compared to the other extended objects. These high deviations in the feature values are due to the hedges' thermal-physical properties. Thus, the hedges display the greatest deviation in thermal radiance throughout its training data since the leaves on the hedges tend to track the availability of solar energy due to the low specific heat of the leaves [3]. Consequently, the features generated for the training data from the thermal images of the hedges captured over diverse environmental conditions, as described in Chap. 2, yield a high deviation among the feature values. The standard deviation of the feature values associated with the brick wall and picket fence are close in value, but lower than the hedges object class and higher than the wood wall object class. Thus, as we see in Fig. 5.3,

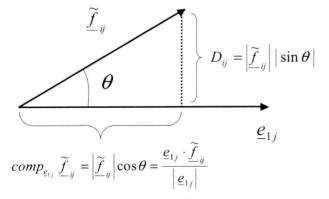

Fig. 5.2 Distance metrics $comp_{\underline{e}_{1j}} \underline{\tilde{f}}_{ij}$ and D_{ij} used to analyze the behavior of each object class's patterns $\underline{\tilde{f}}_{ij}$ about the respective first principal eigenvector $\underline{e}_{1j}$.

166   5 Adaptive Bayesian Classification Model

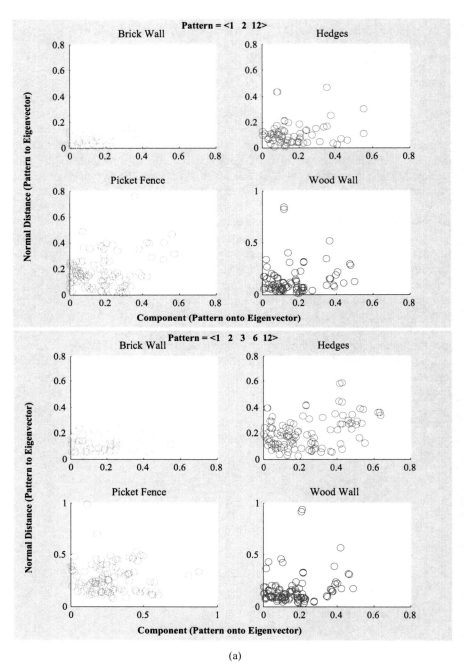

(a)

Fig. 5.3 Extended object distance metric relations for given most favorable feature vector.

5.2 Distance Metrics for Hyperconoidal Clusters

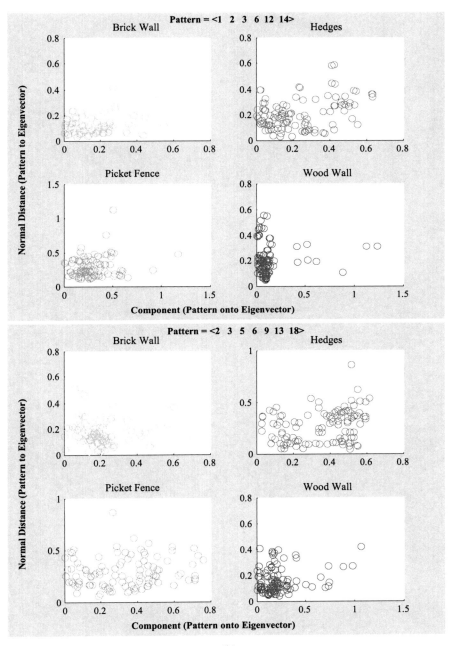

(b)

168   5 Adaptive Bayesian Classification Model

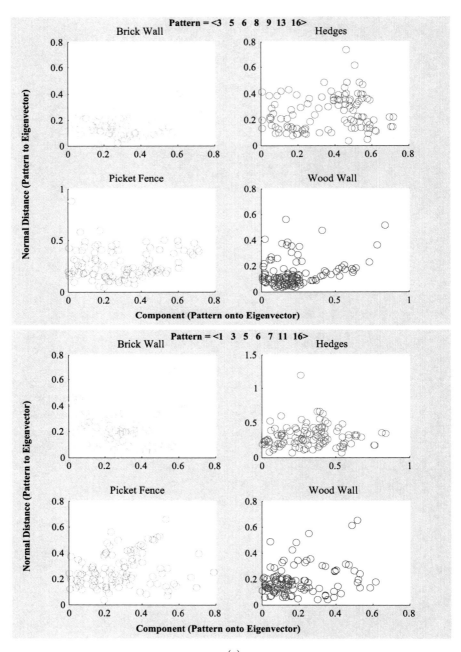

(c)

5.2 Distance Metrics for Hyperconoidal Clusters 169

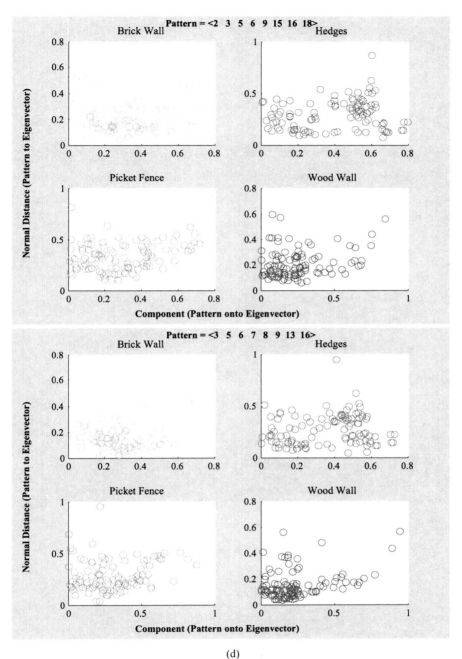

(d)

170   5 Adaptive Bayesian Classification Model

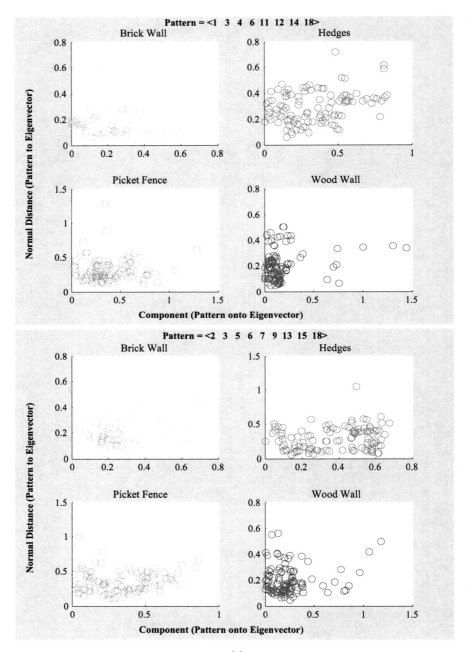

(e)

5.2 Distance Metrics for Hyperconoidal Clusters

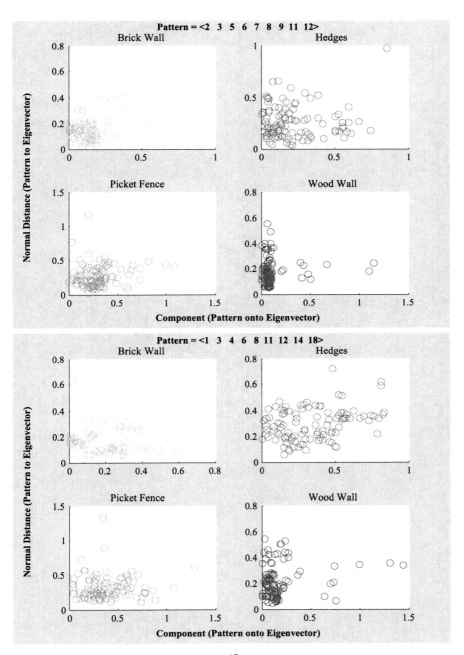

(f)

172   5 Adaptive Bayesian Classification Model

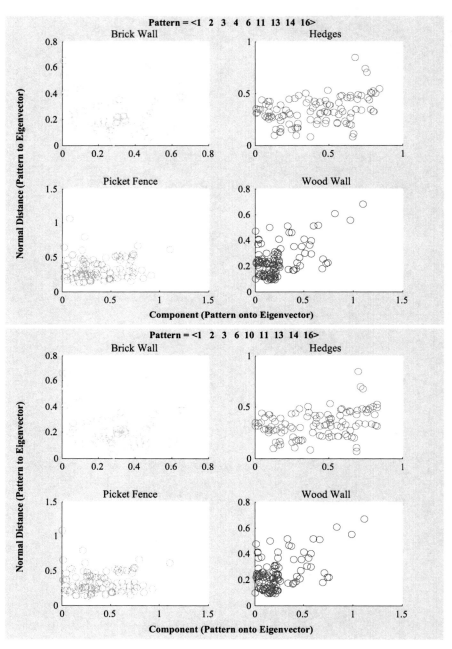

(g)

5.2 Distance Metrics for Hyperconoidal Clusters 173

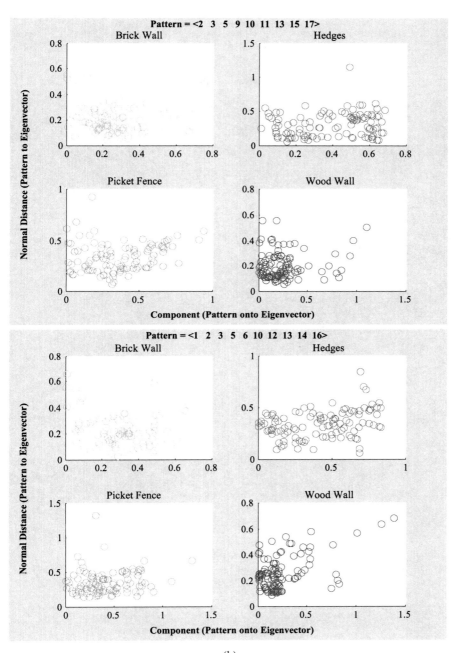

(h)

174 5 Adaptive Bayesian Classification Model

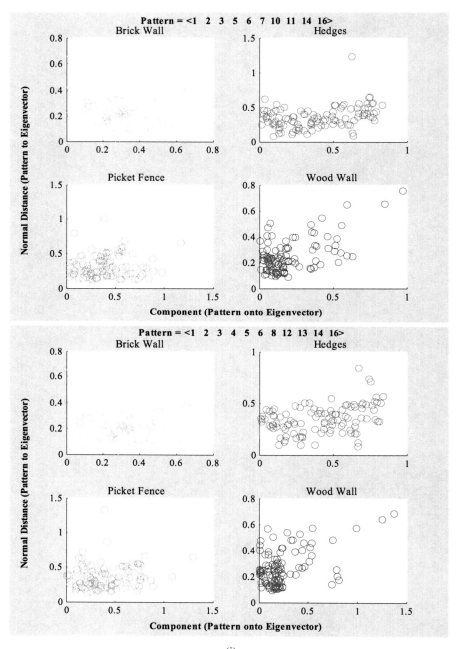

(i)

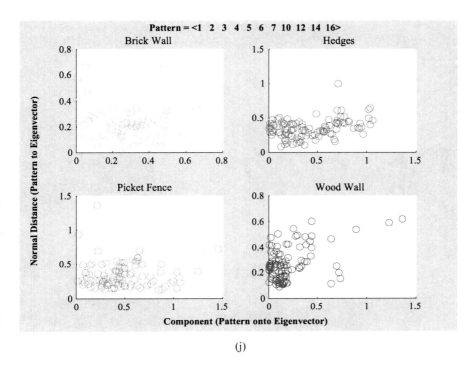

(j)

the relationships between the values of the scalar projections and normal distances for the patterns involving the brick wall and picket fence present approximately the same trends throughout all the feature vectors. On the other hand, the feature values in the training data set for the wood wall object class present the lowest standard deviation compared to the other extended object classes. As we will also discuss in Sect. 5.4.2, the thermal images of the wood walls used in the training data typically had a low thermal radiance and contrast displayed by its surface and reference emitter. The combination of these circumstances contribute to the patterns associated with the wood wall object class displaying a more compact cluster that is closer to the origin in Fig. 5.3 compared to the other extended object classes.

Figures 5.4 (a–q) present the relationships of the distance metrics involving the set of most favorable feature vectors for the compact objects presented in Table 4.17. The numerical labels in each feature vector for the compact objects are defined in Table 4.10. The training data for the compact objects displayed in Table 2.1 is used in this analysis. Analogous to the extended objects, the trends found in the relationships between the distance metrics for the compact objects are attributed to the values of the features for each object class within the training data set. Each object class's feature values depend on its respective material properties, the thermal camera's viewing angle, and the diurnal cycle of solar energy. These combined factors yield

176 5 Adaptive Bayesian Classification Model

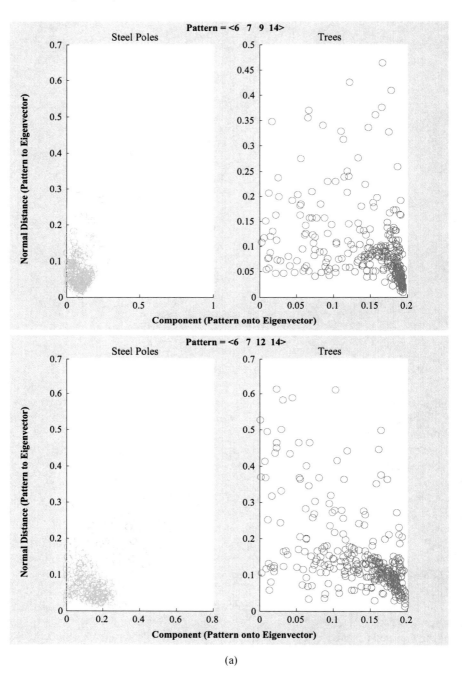

(a)

Fig. 5.4 Compact object distance metric relations for given most favorable feature vector.

5.2 Distance Metrics for Hyperconoidal Clusters 177

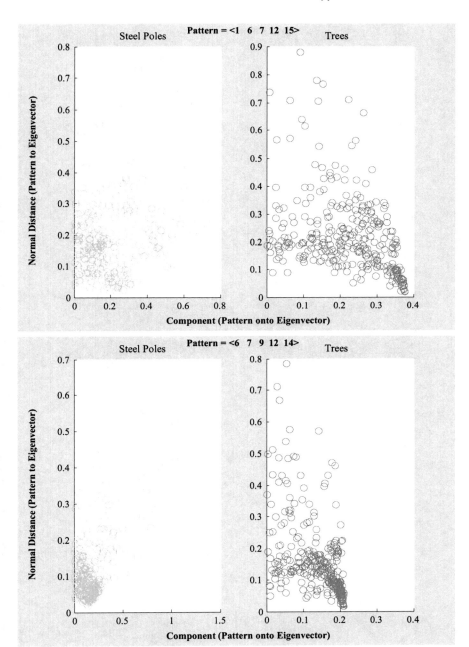

(b)

178 5 Adaptive Bayesian Classification Model

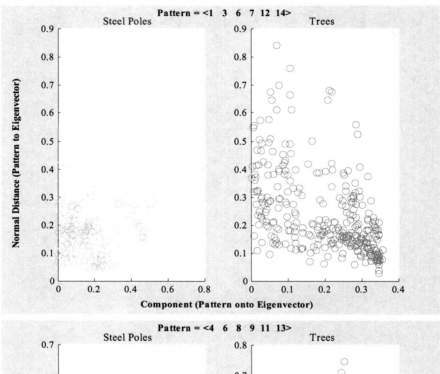
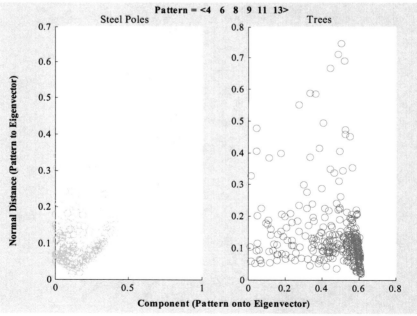

(c)

5.2 Distance Metrics for Hyperconoidal Clusters 179

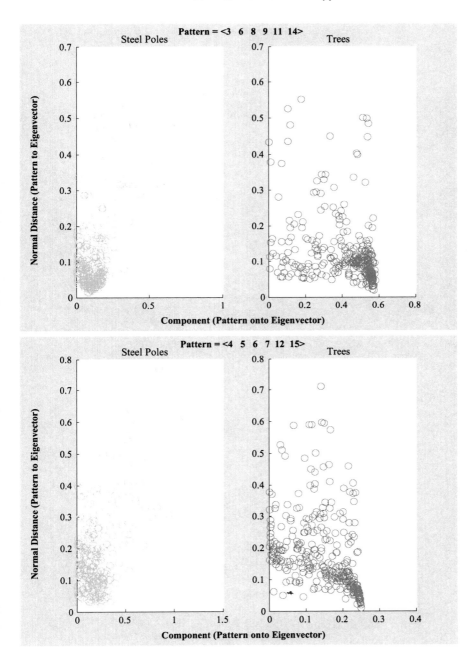

(d)

180 5 Adaptive Bayesian Classification Model

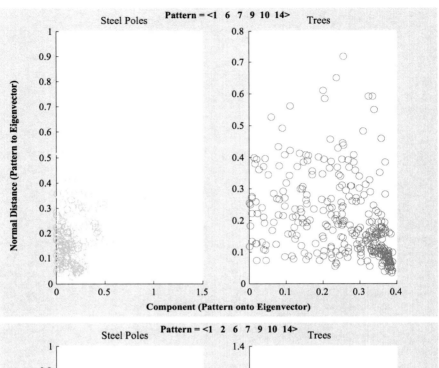

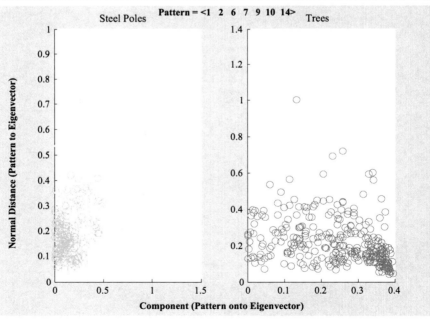

(e)

5.2 Distance Metrics for Hyperconoidal Clusters 181

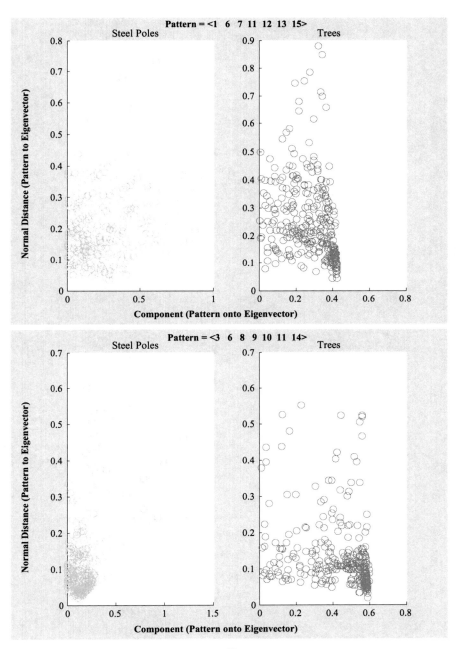

(f)

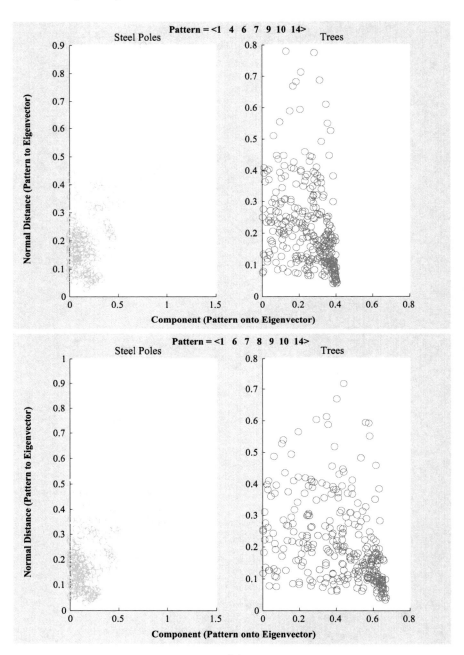

(g)

5.2 Distance Metrics for Hyperconoidal Clusters 183

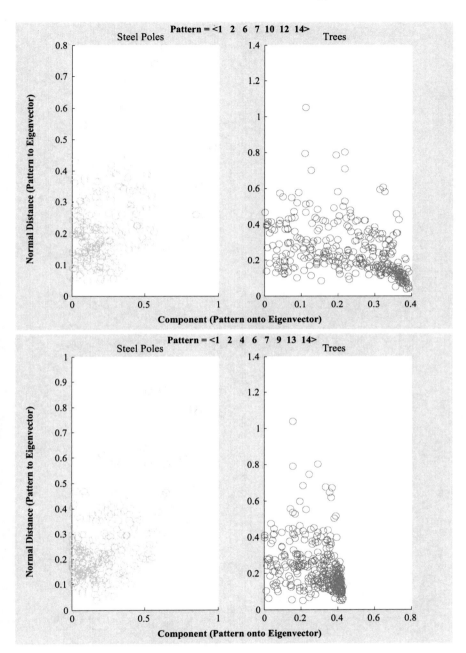

(h)

184   5 Adaptive Bayesian Classification Model

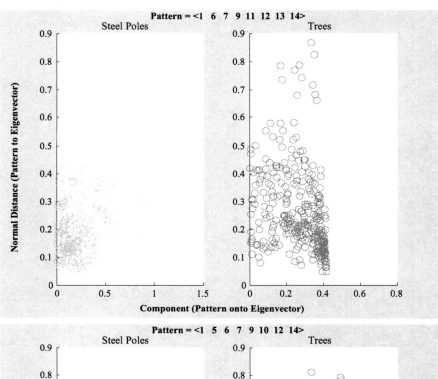

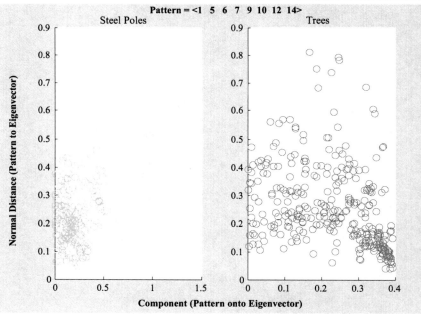

(i)

5.2 Distance Metrics for Hyperconoidal Clusters 185

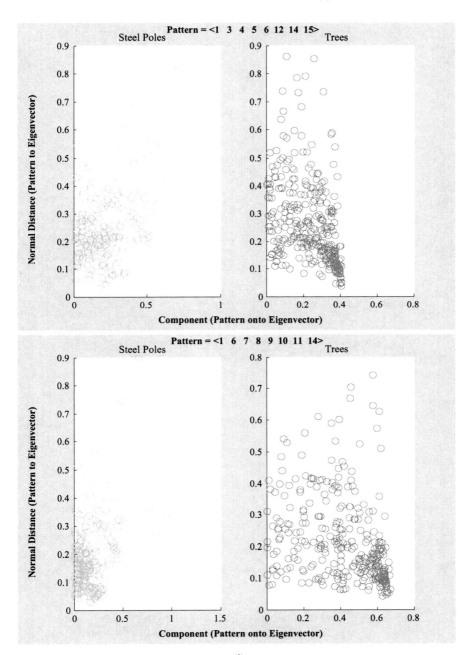

(j)

186   5 Adaptive Bayesian Classification Model

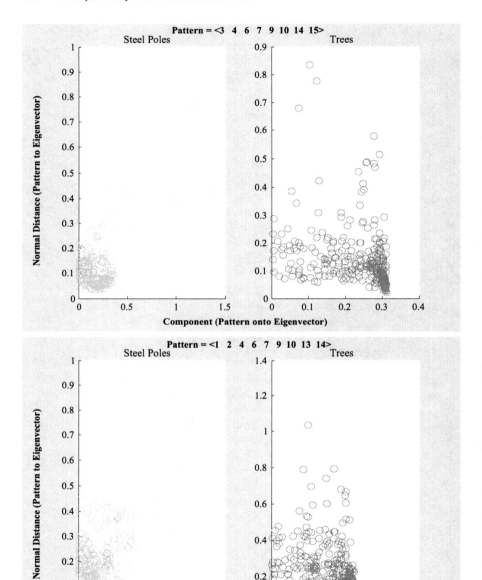

(k)

5.2 Distance Metrics for Hyperconoidal Clusters

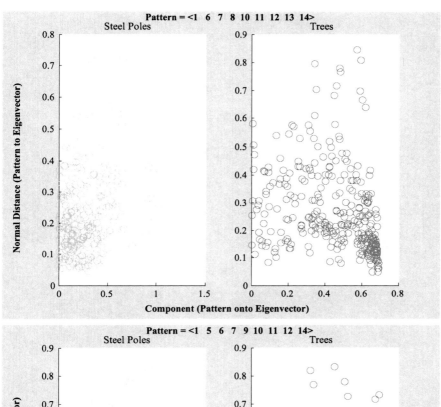

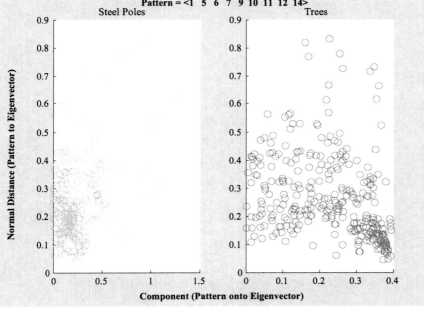

(l)

188 5 Adaptive Bayesian Classification Model

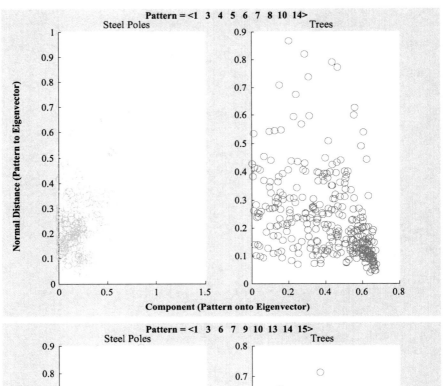
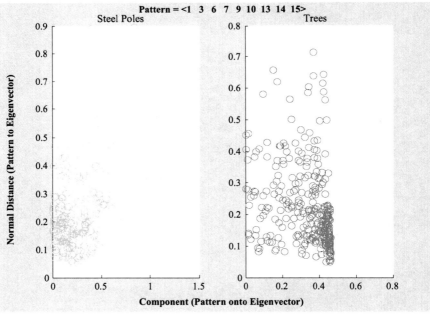

(m)

5.2 Distance Metrics for Hyperconoidal Clusters 189

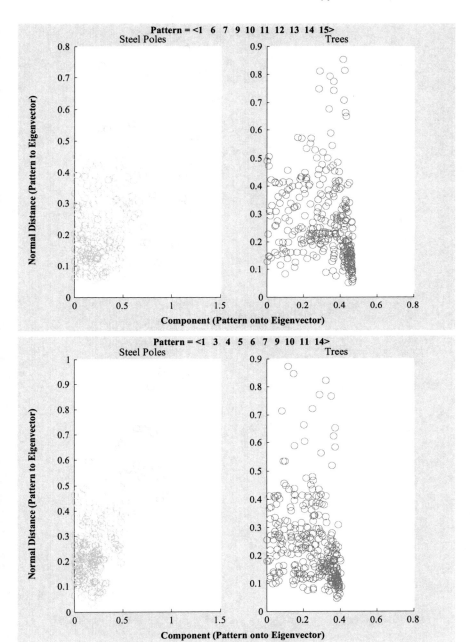

(n)

190 5 Adaptive Bayesian Classification Model

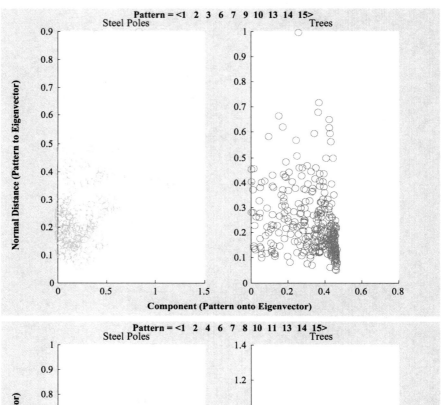

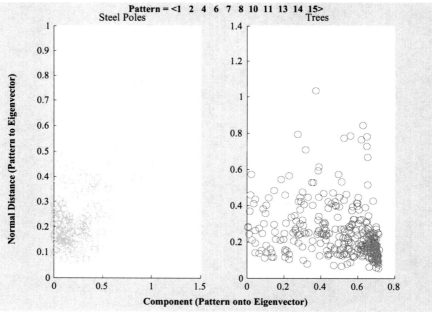

(o)

5.2 Distance Metrics for Hyperconoidal Clusters 191

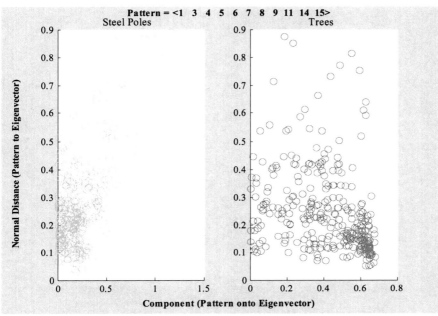

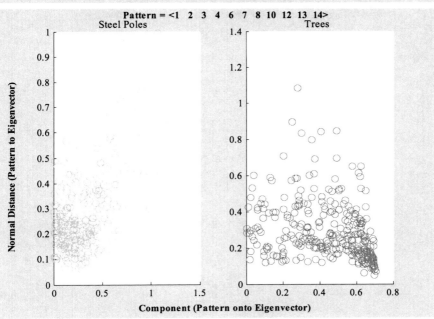

(p)

192   5 Adaptive Bayesian Classification Model

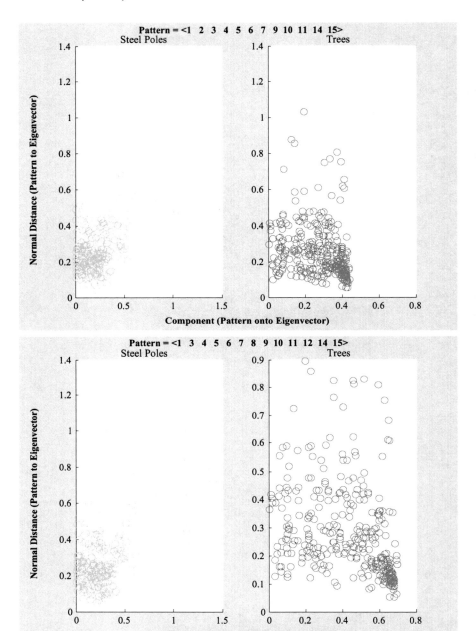

(q)

the trends that we see by each compact object class's patterns in Fig. 5.4. Thus, the standard deviation of the feature values in the steel pole object class's training data set is lower than the tree object class. Furthermore, as we also mention in Sect. 5.4.2, the thermal images of the steel poles used in the training data set typically have a lower thermal contrast displayed on the surface and with the reference emitter compared to the tree object class. These combined factors contribute to the steel pole object class's patterns displaying a more compact cluster that is closer to the origin in Fig. 5.4 compared to the patterns from the tree object class. Consequently, there exists a distinguishing behavior between the patterns for the steel poles and trees.

As we see, the distance metrics given by Eqs. 5.1 and 5.2 play a significant role in the study of n-dimensional patterns that form hyperconoidal clusters. Additionally, these distance metrics give us the ability to "see" regions in the n-dimensional feature space where some object classes may tend to "look alike" and run the risk for misclassification by a classification model. For instance, we will see that the tendency for the majority of the patterns for the wood wall in the extended object category and steel pole in the compact object category to cluster closer to their respective common origins, where minorities of the other object classes' patterns may exist, will lead to a higher error rate for the wood wall and steel pole. Consequently, the common origin for the hyperconoidal clusters of a set of object classes is a region where patterns from the object classes will tend to "look alike." Thus, the closer an object class's patterns are to the common origin of all the hyperconoidal clusters, the higher the risk for misclassification of patterns from that object class by the classification model. Additionally, we can also consider the uniqueness in the behavior of each object class's patterns about their respective first principal eigenvector when assigning a class to an unknown pattern. This is the basis for adaptive Bayesian classifier that we will now present.

5.3 Adaptive Bayesian Classifier Design

Based on our analysis in Sect. 5.2, it appears that the likelihood function used in the posterior probability for classifying an unknown pattern should not only be determined by the unknown pattern's participation in the density distribution of a given object class but also by the unknown pattern's behavior about the first principal eigenvector projecting through the given object class's hyperconoidal cluster. Consequently, we can consider both the density distribution and behavioral characteristics of patterns by deriving a likelihood function that is weighted by a function that involves the normal distance of the unknown pattern to an object class's first principal eigenvector. Additionally, variations of the weighted likelihood function are derived that are adapted to the behavior of the patterns for a given object class. The resulting weighted likelihood function will produce a posterior probability with enhanced discriminating capabilities.

Figure 5.5 presents a zoomed in portion of the hyperconoidal clusters given in Fig. 5.1 with an unknown pattern denoted as a black star in the feature space. Analogous to the behavior of the patterns that we studied in Sect. 5.2, the patterns in the object classes red, blue, and green display unique behaviors about their respective first principal eigenvectors that allow us to distinguish one object class from another. For instance, the patterns in the blue class tend to have a smaller distance to their respective first principal eigenvector compared to the patterns in the red and green classes. The patterns in the green class appear to have larger normal distances from their respective first principal eigenvector compared to the red and blue classes. Consequently, if we computed the normal distance of the unknown pattern (black star) from each object class's first principal eigenvector and combined this information with our knowledge about each object class's density distribution, we could conclude that the characteristics of the black star mostly resemble the blue class. Therefore, the normal distances of the training patterns to the respective first principal eigenvector define the behavior of the given object class.

Let $\tilde{f}$ be an unknown pattern centered for the object classes in an n-dimensional feature space using local PCA. Thus, the unknown pattern is treated as a position vector with an initial position being the common origin for all the hyperconoidal clusters in the n-dimensional feature space. From Eq. 4.13 we have our Bayesian classifier with a KNN density estimation given by

$$P(O_j | \tilde{f}) = \frac{\hat{P}(\tilde{f} | O_j) P(O_j)}{\hat{P}(\tilde{f})} \qquad (5.3)$$

where the likelihood function is defined by the KNN density estimation

$$\hat{P}(\tilde{f} | O_j) = \frac{K_j}{N_j V} \qquad (5.4)$$

and

$$K_j(N_j) = \sqrt{N_j} \qquad (5.5)$$

is a function of the training data in object class O_j as discussed in Chap. 4 and presented by Loftsgaarden and Quesenberry [4]. We will assign equal prior probabilities $P(O_j)$ to the object classes for our analysis throughout this chapter; however, in Chap. 6, we will describe a way to use satellite imagery to assist in establishing prior knowledge used in a bot's area of operation. Consequently, our assignment rule classifies an unknown object to the object class with the largest posterior probability given by Eq. 5.3.

An unknown pattern's normal distance is adapted as a weight on the likelihood function based on the general behavior of the training patterns about each object class's respective first principal eigenvector. For training patterns that tend to

5.3 Adaptive Bayesian Classifier Design

have large normal distances from their respective first principal eigenvector $\underline{e}_{1j}$, such as the green object class in Fig. 5.5, the normal distance D_j for the unknown pattern $\underline{\tilde{f}}$ is adapted as a weighted value on the likelihood function of the object class O_j as a multiplier to obtain

$$\mathcal{L}(\underline{\tilde{f}} \mid O_j) = D_j \cdot P(\underline{\tilde{f}} \mid O_j) \tag{5.6}$$

On the other hand, when the training patterns tend to have smaller normal distances from their respective first principal eigenvector, such as the blue class in Fig. 5.5, the normal distance D_j for the unknown pattern $\underline{\tilde{f}}$ is adapted as a weighted value on the likelihood function of the object class O_j as a divisor to obtain

$$\mathcal{L}(\underline{\tilde{f}} \mid O_j) = \frac{P(\underline{\tilde{f}} \mid O_j)}{D_j} \tag{5.7}$$

Consequently, for an unknown pattern with a large normal distance from a first principal eigenvector, Eq. 5.6 will enhance the likelihood value when the unknown pattern is among a crowd of training patterns from an object that with large normal distances from the same first principal eigenvector. On the other hand, the use of Eq. 5.7 on the unknown pattern (black star) in Fig. 5.5 will enhance the likelihood value of the blue class since the star is among the crowd of blue training patterns. However, the use of Eq. 5.6 on the star for the green class will not yield

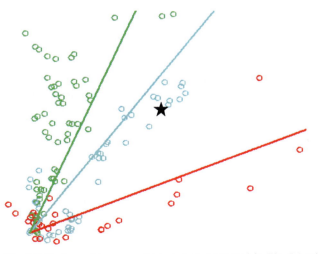

Fig. 5.5 Portion of hyperconoidal clusters presented in Fig. 5.1 with an unknown pattern displayed as the black star in the feature space.

any significant changes to the green class's likelihood value since the star does not exist among a crowd of green training patterns. Such enhancements to the likelihood function will improve the discriminating power of the posterior probability.

We can generalize Eqs. 5.6 and 5.7 to form likelihood functions that are weighted by a distance function $d_j(\tilde{f}, \underline{e}_{1j})$ that takes on various forms involving the normal distances given by Eq. 5.2. For this research, we will consider distance functions $d_j(\tilde{f}, \underline{e}_{1j})$ that are defined by 1, $\frac{1}{D_j}$, $\frac{1}{D_j^2}$, $\frac{1}{\exp(D_j)}$, D_j, D_j^2, and $\exp(D_j)$. Thus, our generalized weighted likelihood function becomes

$$\mathcal{L}(\tilde{f} \mid O_j, d_j) = d_j(\tilde{f}, \underline{e}_{1j}) \cdot P(\tilde{f} \mid O_j) \qquad (5.8)$$

Hence, our weighted KNN density estimation is given by

$$\mathcal{L}(\tilde{f} \mid O_j, d_j) = d_j(\tilde{f}, \underline{e}_{1j}) \cdot \frac{K_j}{N_j V} \qquad (5.9)$$

Therefore, our generalized *adaptive Bayesian classifier* is defined by the posterior probability

$$P(O_j \mid \tilde{f}) = \frac{\mathcal{L}(\tilde{f} \mid O_j, d_j) P(O_j)}{\hat{P}(\tilde{f})} \qquad (5.10)$$

where our unconditional probability is given by

$$\hat{P}(\tilde{f}) = \sum_{j=1}^{J} P(\tilde{f}, O_j)$$
$$= \sum_{j=1}^{J} \mathcal{L}(\tilde{f} \mid O_j, d_j) P(O_j) \qquad (5.11)$$

The novel adaptive Bayesian classifier in Eq. 5.10 puts more weight on the likelihood function when the behavior of an unknown pattern is similar to the patterns of a specific object class. For instance, as previously noted, the unknown pattern (black star) displayed in Fig. 5.5 is located more among the blue class. The adaptive Bayesian classifiers for each class will respond using the weighted likelihood function given by Eq. 5.9. The weighted likelihood function associated with the blue class will have the greatest value since the unknown pattern is among a dense crowd of blue patterns and a close distance to the respective first principal eigenvector like the other blue patterns in the crowd. Consequently, the larger posterior probability associated with the blue class will give us confidence to assign the unknown pattern to the blue object class.

5.4 Adaptive Bayesian Classifier Appraisal

In this section, we will assess the performance of the adaptive Bayesian classifier with the distance function $d_j(\widetilde{\underline{f}},\underline{e}_{1j})$ defined by $1, \frac{1}{D_j}, \frac{1}{D_j^2}, \frac{1}{\exp(D_j)}, D_j, D_j^2$, and $\exp(D_j)$. We will compare these classification results to the performance by the traditional KNN and Parzen classifiers. The classifiers are teamed up with our most favorable feature vectors presented in Chap. 4 and evaluated on the blind data set discussed in Chap. 2. Our analysis will show that our adaptive Bayesian classifiers have the ability to outperform the KNN and Parzen classifiers. Furthermore, we see that some adaptive Bayesian classifiers show exceptional classification performance on a certain object class but do not perform as well on blind patterns from other object classes. This phenomenon is a result of our weighted likelihood functions adapting to the behavior of each object class's patterns about their respective first principal eigenvector. Additionally, we explore why some blind patterns are being misclassified under certain thermal conditions.

5.4.1 Blind Data Performance

We assessed the performance of our adaptive Bayesian classifier given by Eq. 5.10 with the distance function $d_j(\widetilde{\underline{f}},\underline{e}_{1j})$ defined by $1, \frac{1}{D_j}, \frac{1}{D_j^2}, \frac{1}{\exp(D_j)}, D_j, D_j^2$, and $\exp(D_j)$. We compared these classification results to the performance by the traditional KNN and Parzen classifiers presented in Sects. 4.4.2 and 4.4.3, respectively. Each classification model is formed by one of these classifiers and a feature vector from the extended and compact objects displayed in Tables 4.9 and 4.17 presented in Chap. 4. The classification models designed for the extended and compact objects are evaluated on the respective blind data sets for the extended and compact objects discussed in Chap. 2 and presented in Table 2.2.

Tables 5.1 and 5.2 present the average error rates for the adaptive Bayesian, KNN and Parzen classifiers using the most favorable feature vectors and blind data for the extended and compact objects, respectively. The K values for the KNN classifier and h values for the Parzen classifier are presented in blue shaded cells in each table and were derived using the leave-one-out method as discussed in Chap. 4. The numerical labels for the feature vectors of the extended and compact objects are displayed in Tables 4.2 and 4.10, respectively. As we see, the top performers from the adaptive Bayesian classifiers outperform the best models designed from the KNN and Parzen classifiers for both the extended and compact objects. For the extended objects in Table 5.1, the adaptive Bayesian classifier with the distance function $d_j(\widetilde{\underline{f}},\underline{e}_{1j}) = D_j$ and feature vector <1,2,3,4,6,11,13,14,16>

result in the best classification performance with an estimated error rate of 28.26%, while the top performers for both the KNN and Parzen only obtain error rates of approximately 33.70%. For the compact objects in Table 5.2, the adaptive Bayesian classifier with the distance function $d_j(\widetilde{f}, e_{1j})$ defined by 1, D_j, and $\exp(D_j)$ along with the feature vector <1,2,4,6,7,8,10,11,13,14,15> all produce error rates of only 10%, while the KNN and Parzen classifiers both display their best classification performances with error rates of 15%. Therefore, we conclude that the adaptive Bayesian classifier is an appropriate choice for a classification application, such as ours, involving hyperconoidal clusters consisting of patterns in an n-dimensional feature space that are characterized by their behavior about the respective first principal eigenvector.

As expected, the average error rates for the adaptive Bayesian classifier in Tables 5.1 and 5.2 vary with choice of feature vector and distance function $d_j(\widetilde{f}, e_{1j})$. The next question is how these combinations affect the adaptive Bayesian classifier's classification performance on the blind data for each object class within the extended and compact object categories. We analyzed the confusion matrices for the adaptive Bayesian classifier involving every combination of feature vectors and distance functions $d_j(\widetilde{f}, e_{1j})$ in Tables 5.1 and 5.2. Once again, we saw variations in the error rates within each object class with different combinations of feature vectors and distance functions.

Within the extended object class category, the highest error rates consistently occurred with the wood wall object class. The average of the error rates across

Table 5.1 Comparison of average error rates (%) for adaptive Bayesian classifiers with KNN and Parzen classifiers using most favorable feature vectors and blind data for extended objects. The table cells with the lowest average error rates for each classifier are shaded in yellow. The table cell with the overall lowest average error rate is shaded in green.

FEATURE VECTOR	KNN	K Value	Parzen	h Value	ADAPTIVE BAYESIAN CLASSIFIER WITH $d_j(\widetilde{f}, e_{1j})=$						
					1	$\frac{1}{D_j}$	$\frac{1}{D_j^2}$	$\frac{1}{\exp(D_j)}$	D_j	D_j^2	$\exp(D_j)$
1 2 12	68.48	1	68.48	0.0213	56.52	53.26	55.43	55.43	58.70	75.00	55.43
1 2 3 6 12	46.74	1	44.57	0.0297	38.04	38.04	41.30	38.04	40.22	57.61	38.04
1 2 3 6 12 14	47.83	1	45.65	0.0311	38.04	38.04	38.04	38.04	38.04	46.74	36.96
2 3 5 6 9 13 18	45.65	1	44.57	0.0656	40.22	38.04	35.87	39.13	41.30	42.39	40.22
3 5 6 8 9 13 16	35.87	3	40.22	0.0729	40.22	38.04	39.13	40.22	44.57	45.65	42.39
1 3 5 6 7 11 16	42.39	1	41.30	0.0534	41.30	41.30	39.13	40.22	40.22	42.39	41.30
2 3 5 6 9 15 16 18	40.22	1	43.48	0.076	44.57	41.30	39.13	44.57	43.48	45.65	44.57
3 5 6 7 8 9 13 16	33.70	3	40.22	0.0746	41.30	39.13	39.13	42.39	43.48	45.65	41.30
1 3 4 6 11 12 14 18	35.87	1	33.70	0.044	31.52	32.61	32.61	32.61	31.52	32.61	31.52
2 3 5 6 7 9 13 15 18	41.30	1	44.57	0.0691	44.57	43.48	40.22	43.48	44.57	44.57	45.65
2 3 5 6 7 8 9 11 12	47.83	1	46.74	0.0723	41.30	39.13	40.22	40.22	43.48	48.91	40.22
1 3 4 6 8 11 12 14 18	36.96	1	33.70	0.0529	31.52	32.61	31.52	32.61	31.52	33.70	30.43
1 2 2 3 4 6 11 13 14 16	39.13	1	40.22	0.0595	30.43	30.43	29.35	30.43	28.26	32.61	31.52
1 2 3 6 10 11 13 14 16	38.04	1	35.87	0.0463	31.52	29.35	29.35	31.52	30.43	36.96	29.35
2 3 5 9 10 11 13 15 17	39.13	1	44.57	0.0712	44.57	43.48	41.30	44.57	43.48	44.57	45.65
1 2 3 5 6 10 12 13 14 16	38.04	1	35.87	0.0677	33.70	32.61	31.52	32.61	32.61	35.87	33.70
1 2 3 5 6 7 10 11 14 16	46.74	1	46.74	0.0687	32.61	30.43	30.43	30.43	33.70	39.13	32.61
1 2 3 4 5 6 8 12 13 14 16	38.04	1	35.87	0.0641	32.61	31.52	32.61	32.61	33.70	34.78	32.61
1 2 3 4 5 6 7 10 12 14 16	43.48	1	39.13	0.0637	29.35	30.43	30.43	31.52	31.52	34.78	30.43

5.4 Adaptive Bayesian Classifier Appraisal   199

the 133 possible combinations for each object class is: 40.27% for the brick wall, 18.31% for the hedges, 20.17% for the picket fence, and 75.32% for the wood wall. There were nine combinations that resulted in the lowest error rate of 8.70% for the brick wall. Six combinations resulted in the lowest error rate of 8.70% for the hedges. The picket fence had the lowest error rates of 4.35% with four combinations and 8.70% with four other combinations. Six combinations resulted in the lowest error rate of 56.52% for the wood wall. Tables 5.3 (a–d) presents five confusion matrices subjectively selected from the set of combinations of feature vectors and distance functions that resulted in the lowest error rates for each class in the extended object category.

Table 5.2 Comparison of average error rates (%) for adaptive Bayesian classifiers with KNN and Parzen classifiers using most favorable feature vectors and blind data for compact objects. The table cells with the lowest average error rates for each classifier are shaded in yellow. The table cells with the overall lowest average error rate are shaded in green.

(a)

FEATURE VECTOR	KNN	K Value	Parzen	h Value	1	$\frac{1}{D_j}$	$\frac{1}{D_j^2}$	$\frac{1}{\exp(D_j)}$	D_j	D_j^2	$\exp(D_j)$
6 7 9 14	25	9	15	0.0379	17.5	17.5	20	17.5	17.5	20	17.5
6 7 12 14	15	7	20	0.0503	15	15	15	15	15	15	15
1 6 7 12 15	20	7	22.5	0.0515	20	20	20	22.5	15	17.5	15
6 7 9 12 14	20	7	20	0.0466	15	17.5	15	15	15	15	15
1 3 6 7 12 14	22.5	6	20	0.0539	15	22.5	22.5	17.5	15	15	15
4 6 8 9 11 13	25	11	25	0.0667	17.5	17.5	17.5	17.5	22.5	25	17.5
3 6 8 9 11 14	25	8	25	0.061	25	22.5	20	25	27.5	30	25
4 5 6 7 12 15	20	5	20	0.0584	15	17.5	17.5	15	15	17.5	15
1 6 7 9 10 14	22.5	6	22.5	0.0533	15	15	15	15	15	15	15
1 2 6 7 9 10 14	20	3	22.5	0.0549	15	15	15	15	15	15	15
1 6 7 11 12 13 15	22.5	1	22.5	0.0588	15	15	12.5	15	15	20	15
3 6 8 9 10 11 14	25	15	22.5	0.0642	27.5	22.5	20	27.5	27.5	27.5	27.5
1 4 6 7 9 10 14	17.5	6	20	0.0399	17.5	17.5	17.5	17.5	17.5	15	17.5
1 6 7 8 9 10 14	25	3	22.5	0.0653	15	15	15	15	15	15	15
1 2 6 7 10 12 14	20	3	20	0.064	15	15	15	15	15	15	15
1 2 4 6 7 9 13 14	25	12	22.5	0.0688	15	15	17.5	15	15	15	15
1 6 7 9 11 12 13 14	22.5	5	22.5	0.0652	15	15	15	15	15	15	15
1 5 6 7 9 10 12 14	25	5	22.5	0.0647	17.5	17.5	17.5	17.5	15	17.5	17.5
1 3 4 5 6 12 14 15	22.5	7	20	0.0647	22.5	20	20	20	27.5	25	22.5
1 6 7 8 9 10 11 14	22.5	9	22.5	0.065	15	15	15	15	15	15	15
3 4 6 7 9 10 14 15	22.5	7	15	0.0545	12.5	15	15	12.5	12.5	15	12.5
1 2 4 6 7 9 10 13 14	17.5	27	22.5	0.0689	15	15	15	15	15	15	15
1 6 7 8 10 11 12 13 14	22.5	5	22.5	0.0789	15	15	15	15	15	15	15
1 5 6 7 9 10 11 12 14	25	5	22.5	0.0653	22.5	15	25	25	22.5	22.5	22.5
1 3 4 5 6 7 8 10 14	20	6	17.5	0.0742	17.5	17.5	17.5	17.5	17.5	17.5	17.5
1 3 6 7 9 10 13 14 15	22.5	6	20	0.0606	17.5	17.5	17.5	17.5	12.5	15	15

(b)

FEATURE VECTOR	KNN	K Value	Parzen	h Value	1	$\frac{1}{D_j}$	$\frac{1}{D_j^2}$	$\frac{1}{\exp(D_j)}$	D_j	D_j^2	$\exp(D_j)$
1 6 7 9 10 11 12 13 14 15	22.5	5	22.5	0.068	15	15	12.5	15	15	15	15
1 3 4 5 6 7 9 10 11 14	22.5	6	17.5	0.0723	17.5	20	20	17.5	17.5	17.5	17.5
1 2 3 6 7 9 10 13 14 15	25	10	20	0.0624	17.5	17.5	17.5	17.5	15	12.5	15
1 2 4 6 7 8 10 11 13 14 15	22.5	7	17.5	0.0772	10	12.5	15	12.5	10	12.5	10
1 3 4 5 6 7 8 9 11 14 15	22.5	6	17.5	0.0871	17.5	20	20	17.5	17.5	17.5	17.5
1 2 3 4 6 7 8 10 12 13 14	22.5	7	20	0.0852	17.5	17.5	17.5	17.5	17.5	17.5	17.5
1 2 3 4 5 6 7 9 10 11 14 15	22.5	5	17.5	0.0773	17.5	20	20	20	20	20	20
1 3 4 5 6 7 8 9 10 11 12 14 15	22.5	3	20	0.0714	17.5	17.5	17.5	17.5	17.5	17.5	17.5

Table 5.3 Brick wall lowest error rates with respective feature vector and distance function combination displayed in the upper left corner of each confusion matrix.

(a)

Brick Wall

<1,3,4,6,11,12,14,18>, $\frac{1}{D_j^2}$		Actual Object Class			
		Brick Wall	Hedges	Picket Fence	Wood Wall
Assigned Object Class	Brick Wall	21	4	0	12
	Hedges	2	19	1	1
	Picket Fence	0	0	19	7
	Wood Wall	0	0	3	3
Total Objects in Class		23	23	23	23
Errors by Class		2	4	4	20
Error Rate by Class (%)		8.70	17.39	17.39	86.96
Total Errors		30			
Average Error Rate (%)		32.61			

<1,2,3,5,6,10,12,13,14,16>, $\frac{1}{D_j^2}$		Actual Object Class			
		Brick Wall	Hedges	Picket Fence	Wood Wall
Assigned Object Class	Brick Wall	21	4	0	12
	Hedges	2	19	1	1
	Picket Fence	0	0	20	7
	Wood Wall	0	0	2	3
Total Objects in Class		23	23	23	23
Errors by Class		2	4	3	20
Error Rate by Class (%)		8.70	17.39	13.04	86.96
Total Errors		29			
Average Error Rate (%)		31.52			

<1,2,3,4,5,6,8,12,13,14,16>, $\frac{1}{D_j}$		Actual Object Class			
		Brick Wall	Hedges	Picket Fence	Wood Wall
Assigned Object Class	Brick Wall	21	4	0	9
	Hedges	2	19	3	2
	Picket Fence	0	0	19	8
	Wood Wall	0	0	1	4
Total Objects in Class		23	23	23	23
Errors by Class		2	4	4	19
Error Rate by Class (%)		8.70	17.39	17.39	82.61
Total Errors		29			
Average Error Rate (%)		31.52			

5.4 Adaptive Bayesian Classifier Appraisal

<1,2,3,4,5,6,8,12,13,14,16>, $\frac{1}{D_j^2}$

		Actual Object Class			
		Brick Wall	Hedges	Picket Fence	Wood Wall
Assigned Object Class	Brick Wall	21	4	0	12
	Hedges	2	19	3	1
	Picket Fence	0	0	19	7
	Wood Wall	0	0	1	3
	Total Objects in Class	23	23	23	23
	Errors by Class	2	4	4	20
	Error Rate by Class (%)	8.70	17.39	17.39	86.96
	Total Errors	30			
	Average Error Rate (%)	32.61			

<1,2,3,4,5,6,7,10,12,14,16>, $\frac{1}{D_j^2}$

		Actual Object Class			
		Brick Wall	Hedges	Picket Fence	Wood Wall
Assigned Object Class	Brick Wall	21	4	0	10
	Hedges	2	19	1	1
	Picket Fence	0	0	18	6
	Wood Wall	0	0	4	6
	Total Objects in Class	23	23	23	23
	Errors by Class	2	4	5	17
	Error Rate by Class (%)	8.70	17.39	21.74	73.91
	Total Errors	28			
	Average Error Rate (%)	30.43			

Table 5.3 Hedges lowest error rates with respective feature vector and distance function combination displayed in the upper left corner of each confusion matrix.

(b)

Hedges

<1,2,3,6,12>, D_j		Actual Object Class			
		Brick Wall	Hedges	Picket Fence	Wood Wall
Assigned Object Class	Brick Wall	14	2	0	7
	Hedges	9	21	2	6
	Picket Fence	0	0	18	8
	Wood Wall	0	0	3	2
	Total Objects in Class	23	23	23	23
	Errors by Class	9	2	5	21
	Error Rate by Class (%)	39.13	8.70	21.74	91.30
	Total Errors	37			
	Average Error Rate (%)	40.22			

<1,3,4,6,11,12,14,18>, D_j^2		Actual Object Class			
		Brick Wall	Hedges	Picket Fence	Wood Wall
Assigned Object Class	Brick Wall	17	2	0	10
	Hedges	6	21	2	2
	Picket Fence	0	0	18	5
	Wood Wall	0	0	3	6
	Total Objects in Class	23	23	23	23
	Errors by Class	6	2	5	17
	Error Rate by Class (%)	26.09	8.70	21.74	73.91
	Total Errors	30			
	Average Error Rate (%)	32.61			

<1,2,3,5,6,7,10,11,14,16>, 1		Actual Object Class			
		Brick Wall	Hedges	Picket Fence	Wood Wall
Assigned Object Class	Brick Wall	13	2	0	5
	Hedges	9	21	1	4
	Picket Fence	1	0	19	5
	Wood Wall	0	0	3	9
	Total Objects in Class	23	23	23	23
	Errors by Class	10	2	4	14
	Error Rate by Class (%)	43.48	8.70	17.39	60.87
	Total Errors	30			
	Average Error Rate (%)	32.61			

5.4 Adaptive Bayesian Classifier Appraisal

<1,2,3,5,6,7,10,11,14,16>, $\frac{1}{\exp(D_j)}$

		Actual Object Class			
		Brick Wall	Hedges	Picket Fence	Wood Wall
Assigned Object Class	Brick Wall	15	2	0	5
	Hedges	7	21	1	4
	Picket Fence	1	0	19	5
	Wood Wall	0	0	3	9
Total Objects in Class		23	23	23	23
Errors by Class		8	2	4	14
Error Rate by Class (%)		34.78	8.70	17.39	60.87
Total Errors		28			
Average Error Rate (%)		30.43			

<1,2,3,5,6,7,10,11,14,16>, D_j

		Actual Object Class			
		Brick Wall	Hedges	Picket Fence	Wood Wall
Assigned Object Class	Brick Wall	12	2	0	5
	Hedges	10	21	1	4
	Picket Fence	1	0	19	5
	Wood Wall	0	0	3	9
Total Objects in Class		23	23	23	23
Errors by Class		11	2	4	14
Error Rate by Class (%)		47.83	8.70	17.39	60.87
Total Errors		31			
Average Error Rate (%)		33.70			

Table 5.3 Picket fence lowest error rates with respective feature vector and distance function combination displayed in the upper left corner of each confusion matrix.

(c)

Picket Fence

<1,3,5,6,7,11,16>, D_j^2

		Actual Object Class			
		Brick Wall	Hedges	Picket Fence	Wood Wall
Assigned Object Class	Brick Wall	5	2	0	3
	Hedges	15	20	1	6
	Picket Fence	3	1	22	8
	Wood Wall	0	0	0	6
Total Objects in Class		23	23	23	23
Errors by Class		18	3	1	17
Error Rate by Class (%)		78.26	13.04	4.35	73.91
Total Errors		39			
Average Error Rate (%)		42.39			

<1,2,3,4,6,11,13,14,16>, D_j

		Actual Object Class			
		Brick Wall	Hedges	Picket Fence	Wood Wall
Assigned Object Class	Brick Wall	18	3	0	8
	Hedges	5	20	1	3
	Picket Fence	0	0	22	6
	Wood Wall	0	0	0	6
Total Objects in Class		23	23	23	23
Errors by Class		5	3	1	17
Error Rate by Class (%)		21.74	13.04	4.35	73.91
Total Errors		26			
Average Error Rate (%)		28.26			

<1,2,3,4,6,11,13,14,16>, D_j^2

		Actual Object Class			
		Brick Wall	Hedges	Picket Fence	Wood Wall
Assigned Object Class	Brick Wall	16	3	0	8
	Hedges	7	18	1	3
	Picket Fence	0	2	22	6
	Wood Wall	0	0	0	6
Total Objects in Class		23	23	23	23
Errors by Class		7	5	1	17
Error Rate by Class (%)		30.43	21.74	4.35	73.91
Total Errors		30			
Average Error Rate (%)		32.61			

5.4 Adaptive Bayesian Classifier Appraisal

<1,2,3,6,10,11,13,14,16>, D_j

		Actual Object Class			
		Brick Wall	Hedges	Picket Fence	Wood Wall
Assigned Object Class	Brick Wall	17	3	0	8
	Hedges	6	20	1	3
	Picket Fence	0	0	21	6
	Wood Wall	0	0	1	6
	Total Objects in Class	23	23	23	23
	Errors by Class	6	3	2	17
	Error Rate by Class (%)	26.09	13.04	8.70	73.91
	Total Errors	28			
	Average Error Rate (%)	30.43			

<1,2,3,6,10,11,13,14,16>, D_j^2

		Actual Object Class			
		Brick Wall	Hedges	Picket Fence	Wood Wall
Assigned Object Class	Brick Wall	13	2	0	8
	Hedges	10	18	1	3
	Picket Fence	0	3	22	7
	Wood Wall	0	0	0	5
	Total Objects in Class	23	23	23	23
	Errors by Class	10	5	1	18
	Error Rate by Class (%)	43.48	21.74	4.35	78.26
	Total Errors	34			
	Average Error Rate (%)	36.96			

Table 5.3 Wood wall lowest error rates with respective feature vector and distance function combination displayed in the upper left corner of each confusion matrix.

(d)

Wood Wall

<2,3,5,6,7,8,9,11,12>, 1

		Actual Object Class			
		Brick Wall	Hedges	Picket Fence	Wood Wall
Assigned Object Class	Brick Wall	7	3	0	4
	Hedges	16	20	1	5
	Picket Fence	0	0	17	4
	Wood Wall	0	0	5	10
	Total Objects in Class	23	23	23	23
	Errors by Class	16	3	6	13
	Error Rate by Class (%)	69.57	13.04	26.09	56.52
	Total Errors	38			
	Average Error Rate (%)	41.30			

<2,3,5,6,7,8,9,11,12>, $\frac{1}{D_j}$

		Actual Object Class			
		Brick Wall	Hedges	Picket Fence	Wood Wall
Assigned Object Class	Brick Wall	9	3	0	4
	Hedges	14	20	1	5
	Picket Fence	0	0	17	4
	Wood Wall	0	0	5	10
	Total Objects in Class	23	23	23	23
	Errors by Class	14	3	6	13
	Error Rate by Class (%)	60.87	13.04	26.09	56.52
	Total Errors	36			
	Average Error Rate (%)	39.13			

<2,3,5,6,7,8,9,11,12>, $\frac{1}{\exp(D_j)}$

		Actual Object Class			
		Brick Wall	Hedges	Picket Fence	Wood Wall
Assigned Object Class	Brick Wall	8	3	0	4
	Hedges	15	20	1	5
	Picket Fence	0	0	17	4
	Wood Wall	0	0	5	10
	Total Objects in Class	23	23	23	23
	Errors by Class	15	3	6	13
	Error Rate by Class (%)	65.22	13.04	26.09	56.52
	Total Errors	37			
	Average Error Rate (%)	40.22			

5.4 Adaptive Bayesian Classifier Appraisal

<2,3,5,6,7,8,9,11,12>, $\exp(D_j)$

		Actual Object Class			
		Brick Wall	Hedges	Picket Fence	Wood Wall
Assigned Object Class	Brick Wall	7	3	0	4
	Hedges	16	20	1	5
	Picket Fence	0	0	18	4
	Wood Wall	0	0	4	10
Total Objects in Class		23	23	23	23
Errors by Class		16	3	5	13
Error Rate by Class (%)		69.57	13.04	21.74	56.52
Total Errors		37			
Average Error Rate (%)		40.22			

<1,2,3,5,6,7,10,11,14,16>, $\dfrac{1}{D_j}$

		Actual Object Class			
		Brick Wall	Hedges	Picket Fence	Wood Wall
Assigned Object Class	Brick Wall	15	3	0	5
	Hedges	7	20	1	4
	Picket Fence	1	0	19	4
	Wood Wall	0	0	3	10
Total Objects in Class		23	23	23	23
Errors by Class		8	3	4	13
Error Rate by Class (%)		34.78	13.04	17.39	56.52
Total Errors		28			
Average Error Rate (%)		30.43			

Within the compact object class category, the highest error rates consistently occurred with the steel pole object class. The average of the error rates across the 238 possible combinations for each object class is: 29.56% for the steel pole and 4.71% for the tree. Four combinations resulted in the lowest error rate of 20% for the steel pole. The tree had the lowest error rates of 0.00% with 136 combinations of feature vectors and distance functions. Tables 5.4 (a–b) presents four confusion matrices subjectively selected from the set of combinations of feature vectors and distance functions that resulted in the lowest error rates for each class in the compact object category. As displayed in Tables 5.4, the steel pole and tree object classes have the feature vectors for all the chosen combinations and the same distance functions for three of the combinations.

Table 5.4 Steel Pole and Tree lowest error rates with respective feature vector and distance function combination displayed in the upper left corner of each confusion matrix.

(a)

Steel Pole & Tree

<1,2,4,6,7,8,10,11,13,14,15>, 1		Actual Object Class	
		Steel Pole	Tree
Assigned Object Class	Steel Pole	16	0
	Tree	4	20
	Total Objects in Class	20	20
	Errors by Class	4	0
	Error Rate by Class (%)	20	0
	Total Errors	4	
	Average Error Rate (%)	10	

<1,2,4,6,7,8,10,11,13,14,15>, D_j		Actual Object Class	
		Steel Pole	Tree
Assigned Object Class	Steel Pole	16	0
	Tree	4	20
	Total Objects in Class	20	20
	Errors by Class	4	0
	Error Rate by Class (%)	20	0
	Total Errors	4	
	Average Error Rate (%)	10	

<1,2,4,6,7,8,10,11,13,14,15>, $\exp(D_j)$		Actual Object Class	
		Steel Pole	Tree
Assigned Object Class	Steel Pole	16	0
	Tree	4	20
	Total Objects in Class	20	20
	Errors by Class	4	0
	Error Rate by Class (%)	20	0
	Total Errors	4	
	Average Error Rate (%)	10	

Table 5.4 Steel Pole and Tree lowest error rates with respective feature vector and distance function combination displayed in the upper left corner of each confusion matrix.

(b)

Steel Pole

<1,2,4,6,7,8,10,11,13,14,15>, D_j^2		Actual Object Class	
		Steel Pole	Tree
Assigned Object Class	Steel Pole	16	1
	Tree	4	19
	Total Objects in Class	20	20
	Errors by Class	4	1
	Error Rate by Class (%)	20	5
	Total Errors	5	
	Average Error Rate (%)	12.5	

Tree

<1,2,4,6,7,8,10,11,13,14,15>, $\frac{1}{D_j}$		Actual Object Class	
		Steel Pole	Tree
Assigned Object Class	Steel Pole	15	0
	Tree	5	20
	Total Objects in Class	20	20
	Errors by Class	5	0
	Error Rate by Class (%)	25	0
	Total Errors	5	
	Average Error Rate (%)	12.5	

As we see in Tables 5.3 and 5.4, some adaptive Bayesian classifiers show exceptional classification performance on a certain object class but do not perform as well on blind patterns from other object classes. Thus, one classifier may perform exceptionally on specific unknown patterns where another classifier is deficient, and vice versa. In most cases, the classification models presented in Tables 5.3 and 5.4 present better classification results on their respective individual object class than the models' performance on all the classes within their respective extended or compact object category as displayed by the average error rates in Tables 5.1 and 5.2. Consequently, each combination of a feature vector and adaptive Bayesian classifier with the particular distance function displayed in Tables 5.3 and 5.4 forms a model that acts as an *expert* in making classification decisions on patterns from their respective object class. In Sect. 5.5, we will show how a classification model consisting of committees of these experts will further enhance the overall performance.

5.4.2 Analysis of Misclassifications

We next explore why some blind patterns are being misclassified under certain thermal conditions. As discussed in Sect. 5.2, the common origin for the hyperconoidal clusters of a set of object classes is a region where patterns from the object classes will tend to "look alike." Thus, the closer the majority of an object class's patterns are to the common origin of all the hyperconoidal clusters, the higher the risk for misclassification of patterns from that object class by the classification model. We can use the distance metrics given by Eqs. 5.1 and 5.2 to predict what object classes are at risk for misclassification. By relating the scalar project metric in Eq. 5.1 to the normal distance in Eq. 5.2, we saw in Figs. 5.3 and 5.4 that the patterns of the wood wall and steel poles tend to cluster closer to common origin compared to the other object classes in the extended and compact object categories, respectively. Consequently, our classification results in Sect. 5.4.1 verified our predictions since the wood wall and steel pole displayed the highest error rates within their respective object class categories. Now we will go a little deeper "into the bushes" to determine what thermal conditions are required for the patterns from two distinct object classes to "look alike." The analysis consists of finding misclassification trends in both the extended and compact object categories using the confusion matrices resulting from our adaptive Bayesian classifiers and comparing the individual thermal images and feature values of the misclassified objects to those of the respective object classes in the training data set.

5.4.2.1 Misclassifications of Extended Objects

We begin by making inferences on the misclassification of objects within the extended object category. Figure 5.6 displays the visible images and thermal images of a sample of extended objects used in the training data set. The thermal images present the thermal radiance and contrast that are typically found in the scenes for each object class in their respective training data set. The reference emitter (electrical tape) is displayed in each thermal image since it was segmented to generate the Lr feature value as discussed in Chap. 3. Since the extended object training data discussed in Chap. 2 was captured at various viewing angles and times from 15 March to 3 July 2007, there is some deviation in the thermal radiance and contrast for these object classes due to the diurnal cycle of solar energy. Thus, there were times when it was difficult to detect the object and/or distinguish between the object and the reference emitter in the thermal scene. The brick walls used for the training data normally had a low overall thermal radiance and thermal contrast between the brick and the mortar layers. The reference emitter for the brick wall normally had a thermal radiance slightly higher than the brick wall's surface. The hedges normally displayed a good thermal contrast.

5.4 Adaptive Bayesian Classifier Appraisal   211

Fig. 5.6 Visible and thermal images of extended objects from the training data set. The thermal images display the thermal radiance and contrast that are typically found in the scenes for each object class and reference emitters in their respective training data set. (a) brick wall (b) hedges, (c) picket fence, and (d) wood wall.

The reference emitter for the hedges usually had a higher thermal radiance than surface of the hedges. The hedges displayed the greatest deviation in thermal radiance throughout its training data since the leaves on the hedges tend to track the availability of solar energy due to the low specific heat of the leaves [3]. When the hedges are in the shade, a cloud passes, or the sun begins to set, the surface temperature of the hedges stays consistent with the lower ambient temperature and the hedges will display a low thermal radiance in the scene. The picket on the picket fence normally displayed a good thermal contrast with the foreground (in the gaps between the pickets). In the context of this research, we have defined foreground as the region in the scene consisting of objects behind the target of interest and within the thermal camera's field of view. On the other hand, background is defined as the region either in front or to the side of the target consisting of thermal sources that emit thermal energy onto the target's surface. The source emitting this thermal energy may or may not be in the camera's field of view. The reference emitter normally had a higher thermal radiance than the wood surface of the pickets. The thermal radiance and contrast of the wood wall and its reference emitter were normally low, similar to the brick wall.

As noted in Sect. 5.4.1, the brick wall had the second highest average error rate of 40.27% across all combinations of feature vectors and distance functions used by the adaptive Bayesian classifier. The misclassified patterns from the brick wall object class were mainly assigned to the hedges. Figure 5.7 displays the thermal image of one of the misclassified brick walls found in the blind data set that was captured on 24 September 2007 hrs at 1005 hrs. As we see, the high thermal radi-

212   5 Adaptive Bayesian Classification Model

Fig. 5.7 Visible and thermal image of brick wall from the blind data set that was misclassified as a hedge by the adaptive Bayesian classifier. The thermal image was captured on 24 September 2007 at 1005 hrs.

ance of the reference emitter and the high thermal radiance and contrast associated with the blind brick wall in Fig. 5.7 resemble the hedges and reference emitter in Fig. 5.6 (b) more than the brick wall and reference emitter in Fig. 5.6 (a). Consequently, by analyzing the thermal images and feature values, we found that some of the brick wall patterns from the blind data are misclassified as a result of much larger and smaller features values compared to those found in the training data set. The features generated from the misclassified brick wall blind objects that resulted in larger feature values compared to the training data were Lr, $So1$, $En1$, $Co2$, and $En2$. The feature $Er2$ generated from the misclassified brick wall blind objects is smaller in value compared to the brick wall feature values found in the training data set. These results are consistent with the characteristics of our features that we discussed in Chap. 3. In Chap. 3, we noted that $So1$ will take on small values (close to zero) for surfaces with a constant thermal radiance (i.e., gray-level value in the thermal image) and large values (close to unity) when the surface of an object displays large deviations among its gray-level values in the thermal image. Similarly, the feature values for $Co2$, $En1$, and $En2$ will increase for objects with more variations (randomness or complexity) in radiant emissions.

As noted in Sect. 5.4.1, the hedges had the lowest average error rate of 18.31% across all combinations of feature vectors and distance functions used by the adaptive Bayesian classifier. The misclassified hedges were mainly assigned as brick walls. These misclassifications occurred in the thermal images of hedges from the blind data set that presented a low thermal radiance of the reference emitter and low thermal radiance and contrast in the thermal scene associated with the hedges. Figure 5.8 displays the visible and thermal image of one of the hedges that was misclassified as a brick wall. The thermal radiance emitted from the hedges and reference emitter seem to have a stronger resemblance with the thermal image of the brick wall in Fig. 5.6 (a) that is normally found in the training data set. On the other hand, the hedges from the blind data set display a weak resemblance to the thermal radiance of the hedges in Fig. 5.6 (b) that are often found in the training data set. Classifying hedges using a thermal imaging system presents a challenge since the leaves on the hedges tend to track the availability of solar energy due to the low specific heat of the leaves [3]. When the hedges are in the shade, as is the case for the hedges in Fig. 5.7, a cloud passes, or the sun begins to set, the surface temperature of the hedges stays consistent with the lower ambient temperature. Consequently, a low level of solar energy available to this low specific heat object results in less thermal radiation emitted and features that tend to look like those of other objects with a similar thermal scene.

As noted in Sect. 5.4.1, the picket fence had the second lowest average error rate of 20.17% across all combinations of feature vectors and distance functions used by the adaptive Bayesian classifier. The misclassified picket fences from the blind data set were normally assigned as wood walls. Figure 5.9 displays the visible and thermal images of a picket fence from the blind data set that was misclassified as a wood wall. The common characteristics of a picket fence from the blind data that

214 5 Adaptive Bayesian Classification Model

Fig. 5.8 Visible and thermal image of hedges from the blind data set that was misclassified as a brick wall by the adaptive Bayesian Classifier. The thermal image was captured on 15 August 2007 at 1048 hrs.

results in a misclassification as a wood wall are a low thermal radiance emitted from the reference emitter and minimal thermal radiance contrast between the pickets and foreground as we see in Fig. 5.9. Thus, the thermal radiance displayed

Fig. 5.9 Visible and thermal images of a picket fence from the blind data set that was misclassified as a wood wall by the adaptive Bayesian Classifier. The thermal image was captured on 6 October 2007 at 1240 hrs.

by the reference emitter for the picket fence in Fig. 5.9 is similar to the wood wall in Fig. 5.6 (d). In any case, the picket fence will always run the risk of being classified as a wood wall, and vice versa, due to the similar physical and geometrical properties of the two objects.

As noted in Sect. 5.4.1, the wood wall had the highest average error rate of 75.32% across all combinations of feature vectors and distance functions used by the adaptive Bayesian classifier. As we mentioned earlier, the thermal images of the wood walls used in the training data typically had a low thermal radiance and contrast for the wood wall and its reference emitter. As a result, when relating the scalar project metric in Eq. 5.1 to the normal distance in Eq. 5.2, we saw in Figs. 5.3

Fig. 5.10 Visible and thermal images of wood walls from the blind data set that were misclassified by the adaptive Bayesian Classifier. (a) misclassified as a brick wall (captured on 15 August 2007 at 1034 hrs), (b) misclassified as a picket fence (captured on 24 September 2007 at 1029 hrs, same object as in (c) but viewed at normal incidence), (c) misclassified as hedges (captured on 24 September 2007 at 1030 hrs, same object as in (b) but at 45 degrees from normal viewing angle).

that the patterns of the wood wall tend to cluster closer to common origin compared to the other object classes in the extended object category. However, this common origin is a region where the hyperconoidal clusters from all the object classes diverge. Thus, blind objects that are wood walls and have a similar thermal radiance as the wood wall training data in Fig. 5.6 (d) will run the risk of misclassifying in this region where patterns from the object classes will tend to "look alike." On the other hand, if a blind object is a wood wall and its feature values deviate greatly from the norm found in the wood wall's training data, then it will more likely be classified as one of the other object classes.

Now we will go into more detail with the wood walls. Figure 5.10 (a) presents a wood wall from the blind data that was misclassified as a brick wall. Thus, the blind wood wall object and its reference emitter in Fig. 5.10 (a) display a low thermal radiance similar to the thermal images of the brick wall in Fig. 5.6 (a) and wood wall in Fig. 5.6 (d). As we showed in Chap. 4, the thermal features are invariant to the rotation of the given object. Consequently, the thermal contrast resulting from tight fitting, slightly slanted boards forming the blind wood wall in Fig. 5.10 (a) could result in a closer resemblance to the layers of bricks in Fig. 5.6 (a) in feature space.

Three primary conditions that result in a wood wall being misclassified as a picket fence are a high thermal contrast between the wood boards and the foreground in the gaps between the boards, a gap size between the boards that is wider than the typical gaps found in the wood wall's training data, and a reference emitter that produces a higher thermal radiance than the reference emitters with a low thermal radiance in the wood wall's training data. Figure 5.10 (b) displays a thermal image of a blind wood wall object that was misclassified as a picket fence. The thermal features generated from this blind wood wall object would more likely resemble the training data features generated from thermal images of the picket fences captured at 45 degrees from incidence since viewing angles off of normal incidence make the gaps appears smaller.

Three primary conditions that result in a wood wall being misclassified as hedges are a higher thermal radiance emitted by both the surface of the wood wall and reference emitter, high thermal contrast on the surface of the wood boards due to the grains in the wood, and small gaps between the boards of the wood wall. Thus, the combination of these conditions results in thermal features that resemble the complexity (or randomness) associated with hedges. Figure 5.10 (c) displays the same object as Fig. 5.10 (b) capture within one minute apart but at different viewing angles. The wood wall blind object in Fig. 5.10 (c) misclassified as hedges due to these three conditions.

5.4.2.2 Misclassifications of Compact Objects

Figure 5.11 displays the visible images and thermal images of a sample of compact objects used in the training data set. The thermal images present the thermal radiance and contrast that are typically found in the scenes of the training data sets for the steel pole and tree object classes. The reference emitter (electrical tape) is displayed in each thermal image since it was segmented to generate the L_r feature

Fig. 5.11 Visible and thermal images of compact objects from the training data set. The thermal images display the thermal radiance and contrast that are typically found in the scenes for each object class and reference emitters in their respective training data set. Steel poles: (a) brown painted surface, (b) green painted surface, (c) octagon shape, w/ aged brown painted surface. Tree: (d) basswood tree, (e) birch tree, (f) cedar tree.

value as discussed in Chap. 3. Since the compact object training data discussed in Chap. 2 was captured at various viewing angles and times from 15 March to 3 July 2007, there is some deviation in the thermal radiance and contrast for these object classes due to the diurnal cycle of solar energy. Thus, there were times when it was difficult to detect the object and/or distinguish between the object and the reference emitter in the thermal scene. As also discussed in Chap. 3, the steel poles consistently display a relatively constant surface radiance. However, a slight thermal contrast may appear on steel poles with aged painted surfaces that result in flaking of the paint, such as the octagon steel pole in Fig. 5.11 (c). Furthermore, the reference emitter on the surface of the steel poles is normally difficult to distinguish from the steel poles' surfaces since the emissivity of the electrical tape ($\varepsilon \sim 0.97$) is about the same as emissivity of the steel poles' surfaces ($\varepsilon \sim 0.92$–0.96 at $75.2°F$ depending on the type of paint) [3]. The trees' surfaces typically displayed a high thermal contrast due to the large variations in the radiance from the bark patterns. However, the birch tree's surface usually presented the lowest thermal contrast, compared to the other trees, due to the less rough characteristics of its bark. The reference emitter attached to the trees' surfaces normally displayed a higher thermal radiance than the trees' surfaces.

As noted in Sect. 5.4.1, the steel pole had the highest average error rate of 29.56% across all combinations of feature vectors and distance functions used by the adaptive Bayesian classifier. When relating the scalar project metric in Eq. 5.1 to the normal distance in Eq. 5.2, we saw in Figs. 5.4 that the patterns of the steel pole tend to cluster closer to common origin compared to the patterns of the tree object class. As previously discussed, this common origin is a region where the hyperconoidal clusters from all the object classes diverge. Thus, blind objects that are steel poles and have a similar thermal radiance and contrast as the steel pole training data in Fig. 5.6 (a–c) will run the risk of misclassifying in this region where patterns from the object classes will tend to "look alike." Additionally, if a blind object is a steel pole and its feature values deviate greatly from the norm found in the steel pole's training data, then it may be misclassified as a tree. Figure 5.12 displays the visible and thermal images of a steel pole captured on 5 November 2007 at 1428 hrs for the blind data that consistently misclassified as a tree. This steel pole used for the blind data was an unpainted, lightly oxidized surface with an emissivity of approximately $\varepsilon \sim 0.80$ at $77°F$ [3]. Consequently, the electrical tape reference emitter, with an emissivity of approximately $\varepsilon \sim 0.97$, emits a higher thermal radiance compared to the surface of the steel pole. Furthermore, the oxidized surface of the steel pole results in a thermal contrast that is seen in the steel pole's thermal image. The combination of the thermal contrast on the surface of the steel pole and higher emission of thermal radiation by the reference emitter results in a thermal scene similar to the trees in the training data set and misclassification by the adaptive Bayesian classifiers.

As noted in Sect. 5.4.1, the tree object class had the highest average error rate of 4.71% across all combinations of feature vectors and distance functions used by the adaptive Bayesian classifier. Figure 5.13 displays the visible and thermal image of a tree from the blind data set that misclassified as a steel pole. The obvious conditions that will result in a misclassification of a tree as a steel pole are a low thermal contrast on the surface of the tree and thermal radiant emission from the reference emitter that is similar to the tree's surface as displayed in Fig. 5.13. Consequently, the thermal image of the tree in Fig. 5.13 has characteristics that are similar to the steel poles in Fig. 5.11 (a–c).

Fig. 5.12 Visible and thermal images of a steel pole from the blind data set that was misclassified as a tree by the adaptive Bayesian Classifier. The thermal image was captured on 5 November 2007 at 1428 hrs.

5.4 Adaptive Bayesian Classifier Appraisal 221

Fig. 5.13 Visible and thermal images of a tree from the blind data set that was misclassified as a steel pole by the adaptive Bayesian Classifier. The thermal image was captured on 18 September 2007 at 1407 hrs.

5.4.2.3 Misclassifications Discussion

The correct classification of a blind object was independent of the geographical location of the object. For instance, the adaptive Bayesian classifier was just as successful in correctly classifying a picket fence in Buffalo, New York, as it was in York County, Virginia. The two primary factors that contributed to the misclassification of the blind objects were a lack of representative training data and the effects of the diurnal cycle of solar energy. Thus, some misclassifications could be eliminated by expanding the range of features in the training data set by capturing a more representative set of thermal images. However, in most cases a lack of a thermal signature from an object due to the diurnal cycle of solar energy will continue to result in feature values from different object classes looking alike. As discussed in Chap. 4, the phenomenon primarily responsible for a target and the surrounding surfaces having approximately the same level of thermal radiant emissions is known as thermal crossover [3]. Thermal crossover results in minimal thermal contrast between the surfaces of objects and the surrounding environment within the thermal infrared camera's field of view. Consequently, thermal images of objects captured during thermal crossover run the risk of producing features that the bot will think look like features from other object classes. In Chap. 6, we will discuss how these periods of thermal crossover could result in a limitation to our ability to classify non-heat generating objects in an outdoor environment using a thermal infrared imaging sensor. We will also present a method that integrates a thermal contrast threshold rule into the detection phase of the classification process that requires a minimum amount of contrast in the scene to use the thermal infrared imaging sensor. If the rule is not satisfied, the autonomous robot must reject the use of the thermal imaging sensor and rely on other sensors such as ultrasound to assist in classifying the target.

Another observation from our analysis is that in some cases the misclassification of a blind object was associated with either a low posterior probability or a posterior that was close in value to another posterior for an assignment to a different object class. Although the posterior probabilities provide a degree of certainty in the bot's ability to correctly classify an unknown object, these two situations may increase risk of misclassification and decrease our confidence in the bot's classification decision. We can gain more confidence in the bot's decisions by integrating certain rules into the classification model that will require the bot to capture another thermal image of an unknown object if these rules are not satisfied. For instance, if the classification model's resulting posterior probability for assigning an unknown pattern to an object class does not satisfy a specific threshold, then the classification is rejected and the bot is required to capture another image, perhaps at another viewing angle, for class assignment. We will present these types of rules with our novel adaptive Bayesian classification model in Sect. 5.5. The tendency for an object to "look like" another object under certain thermal conditions (other than thermal crossover) presents a degree of vagueness that may call for the integration of fuzzy logic into the classification model. Additionally, we could integrate other sensors into the autonomous robotic system by designing a multi-sensor data fusion architecture where the use of multiple sensors complements the overall performance of the classification model. We will discuss our plans for future research involving the integration of fuzzy logic into our classification model and designing a multi-sensor classification model in Chap. 6.

5.5 Adaptive Bayesian Classification Model Design

We now present the design of our *Adaptive Bayesian Classification Model*. In Sect. 5.4.1, we saw that some adaptive Bayesian classifiers show exceptional classification performance on a certain object class but do not perform as well on blind patterns from other object classes. Thus, one classifier may perform exceptionally on unknown patterns from a specific object class where another classifier is deficient, and vice versa. Consequently, each combination of a feature vector and adaptive Bayesian classifier with the particular distance function displayed in Tables 5.3 and 5.4 forms a model that acts as an *expert* in making classification decisions on unknown patterns from the respective object class. By forming a committee of these experts of a specific object class, we should have a model with improved classification performance and confidence in deciding whether an unknown pattern belongs to the respective object class. With multiple committees, each consisting of experts of a specific object class, one committee of experts will perform exceptionally on specific unknown patterns where another classifier is deficient, and vice versa. By combining each committee of experts into one classification model, we are able to exploit the expertise of each committee and complement the overall performance of the classification model. We can increase the confidence level in our model's classification decisions by integrating the *dynamical window* technique presented in Chap. 4 that lets each committee of experts decide on class assignment by considering information collected from multiple window sizes of the thermal image of an object. Additionally, we can integrate rules to improve the accuracy of class assignments and prevent voting ties by the committees. Included are rules that will require the bot to reject class assignments if a posterior probability is below a given threshold or too close to another committee's posterior probability. This will prevent decisions on class assignments during these high-risk situations. Rejections of a class assignment will require the bot to capture another thermal image of the unknown object for classification, perhaps at another viewing angle. We will seek to choose threshold values that minimize both the error rate and number of rejections of class assignments. This is the cornerstone of our Adaptive Bayesian Classification Model.

The concept behind our Adaptive Bayesian Classification Model resides in the topic of *combining classifiers*. There are many strategies for combining classifiers [5, 6, 7]. Analogous to what is found for single classifiers, there is no universal combination of classifiers. The combination of classifiers is chosen based on how well it performs for a specific pattern classification application. Thus, the *No Free Lunch Theorem* discussed in Chap. 4 prevails again. Consequently, the Adaptive

224   5 Adaptive Bayesian Classification Model

Bayesian Classification Model is an appropriate choice for any classification application, such as ours, involving hyperconoidal clusters consisting of patterns in an *n*-dimensional feature space that are characterized by their behavior about their respective first principal eigenvector.

Figure 5.14 presents our algorithm for the Adaptive Bayesian Classification Model designed to assign classes to objects from the extended object category. As

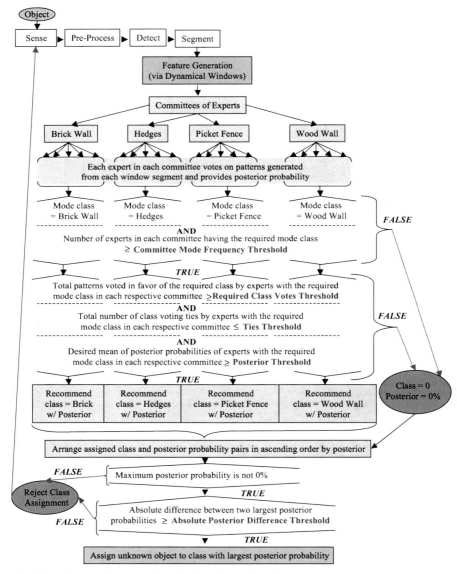

Fig. 5.14 Adaptive Bayesian Classification Model Algorithm.

we will show, this algorithm can be easily modified to support the compact object category. The algorithm begins with the thermal infrared imaging camera receiving thermal radiation emitted from objects within the camera's field of view as described in Chap. 2. The thermal image of the scene is pre-processed as discussed in Chap. 2. After pre-processing, existing algorithms are used to detect and segment an unknown object in the thermal image. The curvature algorithm, introduced in Chap. 3, is used to distinguish and separate extended objects from compact objects. The bot will then use the Adaptive Bayesian Classification Model to assign a class to the unknown object.

Once the unknown object is categorized as either an extended or compact object, the respective set of thermal features are generated from window segments of the object's thermal image that vary in size by a technique that we refer to as *dynamical windows*. Dynamical windows increases our confidence level in our model's final classification decision since the technique lets the bot make a decision on the class assignment of an unknown object by interpreting information collected from multiple window sizes of the thermal image of the object. This technique is analogous to how a human would perhaps study an object at varying fields of view to make a class assignment. In Sect. 4.6.2, we saw that generating thermal features from 100 window segments of an extended object's thermal image that decrease in size will result in posterior probabilities computed by a Bayesian classification model that generally display minimal variations until about the 80th window size index. Thus, the posterior probabilities produced by the classification model generally became sensitive to the smaller window segments with an index greater than 80, resulting in inconsistent posterior probabilities and class assignments. Consequently, we will apply the dynamical window technique by generating thermal features from 80 window segments of decreasing size. For the extended objects' thermal features displayed in Table 4.2, the micro features generated from a segment of the object's surface and meteorological features will remain constant during the classification of the given unknown object. However, the values for the macro features will be computed for each window size. For the compact objects' thermal features displayed in Table 4.10, the micro features Lr and Lb will remain constant during the classification of the given unknown object. However, the micro features Lo, Lor, Lob, and Eo and macro features will be computed for each window size since Lo and the macro features are generated from the same center segment of a given compact object.

A committee of experts is formed for each object class within both the extended and compact object categories. As mentioned previously, each expert consists of a feature vector and adaptive Bayesian classifier with a particular distance function that performs exceptionally on classifying unknown patterns from a specific object class. Table 5.3 displays the experts for the extended objects. For the extended object category, each object class consists of five experts that form a committee of experts. As we see in Table 5.4, the steel pole and tree compact object classes each have four experts in their respective committee. The selection of the number and types of experts is subjective; however, the goal should always be to select the experts for each object class that result in exceptional classification performance.

Additional research is required to determine the most favorable number of experts in each committee. Each expert in each committee votes on the 80 patterns generated from each dynamical window segment of the unknown object's thermal image by assigning a class and posterior probability.

The next phase in the algorithm consists of the first set of classification rules. Since each committee is an expert in classifying unknown patterns from a specific object class, a majority vote by the experts in a given committee of an unknown object being assigned to their respective object class would give us some confidence that the committee of experts is correct. Consequently, the first rule requires that the majority (or mode) of votes of the experts in a given committee be in favor of their respective object class, known as the *mode class*. Furthermore, the number of experts in each committee having the respective object class as their mode must be greater than or equal to a *Committee Mode Frequency Threshold*. The Committee Mode Frequency Threshold is a value from the set $\{1, 2, 3,...,n\}$ where n is the number of experts in each committee. For instance, a mode class equal to brick wall and a Committee Mode Frequency Threshold = 3 implies that the mode of the votes for the committee must be in favor of the brick wall and at least 3 of the 5 experts in the committee must have the brick wall as their mode when voting on the unknown pattern. If these two rules are not satisfied for a given committee, the committee assigns a class label of 0 to the unknown object with a posterior probability of 0%. If the rules are satisfied, the given committee's voting information moves on to the next set of rules.

The next set of rules applies to those experts with the required mode class in each respective committee. The first rule is that the total number of patterns voted in favor of the required class by the experts with the required mode class in each committee must be greater than or equal to a *Required Class Votes Threshold*. The chosen Required Class Votes Threshold is a number no greater than the product of the number of experts in a given committee and number of dynamical window segments (i.e., $5*80 = 400$ for our extended object application). The choice for the Required Class Votes Threshold is associated with the Committee Mode Frequency Threshold. For instance, with the extended objects, if Committee Mode Frequency Threshold = 2 and Required Class Votes Threshold = 400 are selected, the rule involving Committee Mode Frequency Threshold may be satisfied; however, it is very possible that the rule involving the Required Class Votes Threshold may not be satisfied. The second rule is that the total number of class voting ties by the experts with the required mode class in each respective committee must be less than or equal to the *Ties Threshold*. By expecting the ideal situation, where there are no class voting ties by each expert on the total number of patterns produced by the dynamical window, the *Ties Threshold* = 0. The third rule is that the desired mean of the posterior probabilities of the experts with the required mode class in each respective committee is greater than or equal to a *Posterior Threshold*. The Posterior Threshold is chosen based on the degree of confidence desired for each committee's recommendation for a class assignment of the unknown object. If a committee satisfies these three rules, then its mode class and the associated mean of the posterior probabilities of the experts with the required mode class is provided as the committee's recommendation for a class assignment of the unknown object. Otherwise, if the three rules are not satisfied, the committee assigns a class label of 0 to the unknown object with a posterior probability of 0%.

The final phase in the algorithm for the Adaptive Bayesian Classification Model involves all the committees to present their recommendations for the class assignment of the unknown object. The recommended assigned class and their respective posterior probability from each committee are arranged in ascending order by the posterior probabilities. If the maximum posterior probability among all the committees is not 0%, then the recommended class assignment information from each committee is sent to the final decision rule. Otherwise, if the maximum posterior probability is 0%, then all the committees recommended a class label of 0 to the unknown object with a posterior probability of 0% and the class assignment is rejected by the model. The final decision rule is that the absolute difference between the two largest posterior probabilities is greater than or equal to an *Absolute Posterior Difference Threshold*. The Absolute Posterior Difference Threshold will prevent high-risk situations of assigning a class to an unknown object when two committees made different class assignment decisions but have a small difference in their posterior values. This threshold will also eliminate ties when two committees vote on different class assignments but they have the same posterior probability values. If the rule involving the Absolute Posterior Difference Threshold is not satisfied, then the recommended class assignment is rejected by the model. Otherwise, if the Absolute Posterior Difference Threshold is satisfied, the unknown object is assigned to the class with the largest posterior probability.

Rejections of a class assignment will require the bot to capture another thermal image of the unknown object for classification, perhaps at another viewing angle. Consequently, the Adaptive Bayesian Classification Model would be appropriate for autonomous robotic systems that capture continuous frames. If the class assignment is accepted by the Adaptive Bayesian Classification Model, the bot will use this classification output to decide on the next required action in the intelligence algorithm [*report the object* and/or (*if the object is a hedge, go through the object* or *if the object is a brick wall, go around the object*)].

5.6 Adaptive Bayesian Classification Model Application

In this section we will assess the performance of the Adaptive Bayesian Classification Model presented in Sect. 5.5 on the extended and compact blind data displayed in Table 2.2. We will also evaluate the classification model's response when confronted with the following additional blind objects that include objects outside the classes in the training data sets: brick wall with moss on the surface, concrete wall, bush, gravel pile, steel picket fence, wood bench, wood wall of a storage shed, square steel pole, aluminum pole for a dryer vent, concrete pole, knotty tree, telephone pole, 4×4 wood pole, and pumpkin.

5.6.1 Performance on Blind Data (with Classes = Training Set)

The performance of the Adaptive Bayesian Classification Model on the blind data in Table 2.2 was analyzed using various combinations of values for the model's thresholds. As discussed in Sect. 2.3, the blind data presented in Table 2.2 consisted of the same classes and were captured at the same viewing angles as the training data but were not the same objects. The blind data was classified by the Adaptive Bayesian Classification Model using a login node on the DoD High Performance Computing Modernization Program system at the Army Research Laboratory Major Shared Resource Center that included 8 GB of memory at a processor frequency of 3.6 GHz. The model required approximately 4.45 minutes to make a decision regarding the class assignment of each object in the extended object category consisting of the brick wall, hedges, picket fence, and wood wall object classes. The model required approximately 1.16 minutes to classify each object in the compact object category consisting of the steel pole and tree object classes. Tables 5.5 and 5.6 provide the confusion matrices of the Adaptive Bayesian Classification Model with different combinations of threshold values for the extended and compact object categories, respectively. The confusion matrices include the number of objects that were rejected by the Adaptive Bayesian Classification Model due to the rules in the model not being satisfied. The model's rejections of class assignments do not count toward the error rates.

Table 5.5 Confusion matrices of the Adaptive Bayesian Classification Model with various threshold values for the extended objects. Fixed threshold values are noted in the upper left corner. Threshold with a varied value is noted at the upper left corner of each matrix. Thresholds highlighted in green colored text are selected as most favorable for the Adaptive Bayesian Classification Model applied to the extended objects.

(a)

Committee Mode Frequency Threshold = 4
Required Class Votes Threshold = 250
Ties Threshold = 0
Absolute Posterior Difference Threshold = 0.10

Posterior Threshold = 0.6

		Actual Object Class			
		Brick Wall	Hedges	Picket Fence	Wood Wall
Assigned Object Class	Brick Wall	18	3	0	6
	Hedges	3	19	0	1
	Picket Fence	0	0	18	5
	Wood Wall	0	0	3	5
Rejections by Class		2	1	2	6
Total Objects in Class		23	23	23	23
Errors by Class		3	3	3	12
Error Rate by Class (%)		14.29	13.64	14.29	70.59
Total Errors		21			
Average Error Rate (%)		28.20			
Total Rejections		11			

5.6 Adaptive Bayesian Classification Model Application

Posterior Threshold = 0.7

		Actual Object Class			
		Brick Wall	Hedges	Picket Fence	Wood Wall
Assigned Object Class	Brick Wall	17	3	0	6
	Hedges	3	20	0	1
	Picket Fence	0	0	17	5
	Wood Wall	0	0	3	5
Rejections by Class		3	0	3	6
Total Objects in Class		23	23	23	23
Errors by Class		3	3	3	12
Error Rate by Class (%)		15.00	13.04	15.00	70.59
Total Errors		21			
Average Error Rate (%)		28.41			
Total Rejections		12			

Posterior Threshold = 0.8

		Actual Object Class			
		Brick Wall	Hedges	Picket Fence	Wood Wall
Assigned Object Class	Brick Wall	17	3	0	6
	Hedges	3	14	0	1
	Picket Fence	0	0	16	4
	Wood Wall	0	0	2	4
Rejections by Class		3	6	5	8
Total Objects in Class		23	23	23	23
Errors by Class		3	3	2	11
Error Rate by Class (%)		15.00	17.65	11.11	73.33
Total Errors		19			
Average Error Rate (%)		29.27			
Total Rejections		22			

Posterior Threshold = 0.9

		Actual Object Class			
		Brick Wall	Hedges	Picket Fence	Wood Wall
Assigned Object Class	Brick Wall	15	3	0	6
	Hedges	1	13	0	1
	Picket Fence	0	0	15	3
	Wood Wall	0	0	1	1
Rejections by Class		7	7	7	12
Total Objects in Class		23	23	23	23
Errors by Class		1	3	1	10
Error Rate by Class (%)		6.25	18.75	6.25	90.91
Total Errors		15			
Average Error Rate (%)		30.54			
Total Rejections		33			

Table 5.5 Confusion matrices of the Adaptive Bayesian Classification Model with various threshold values for the extended objects. Fixed threshold values are noted in the upper left corner. Threshold with a varied value is noted at the upper left corner of each matrix. Thresholds highlighted in green colored text are selected as most favorable for the Adaptive Bayesian Classification Model applied to the extended objects.

(b)

Committee Mode Frequency Threshold = 4
Ties Threshold = 0
Posterior Threshold = 0.6
Absolute Posterior Difference Threshold = 0.10

Required Class Votes Threshold = 1

		Actual Object Class			
		Brick Wall	Hedges	Picket Fence	Wood Wall
Assigned Object Class	Brick Wall	18	3	0	6
	Hedges	3	19	0	1
	Picket Fence	0	0	18	5
	Wood Wall	0	0	3	5
Rejections by Class		2	1	2	6
Total Objects in Class		23	23	23	23
Errors by Class		3	3	3	12
Error Rate by Class (%)		14.29	13.64	14.29	70.59
Total Errors		21			
Average Error Rate (%)		28.20			
Total Rejections		11			

Required Class Votes Threshold = 50

		Actual Object Class			
		Brick Wall	Hedges	Picket Fence	Wood Wall
Assigned Object Class	Brick Wall	18	3	0	6
	Hedges	3	19	0	1
	Picket Fence	0	0	18	5
	Wood Wall	0	0	3	5
Rejections by Class		2	1	2	6
Total Objects in Class		23	23	23	23
Errors by Class		3	3	3	12
Error Rate by Class (%)		14.29	13.64	14.29	70.59
Total Errors		21			
Average Error Rate (%)		28.20			
Total Rejections		11			

5.6 Adaptive Bayesian Classification Model Application

Required Class Votes Threshold = 100

		Actual Object Class			
		Brick Wall	Hedges	Picket Fence	Wood Wall
Assigned Object Class	Brick Wall	18	3	0	6
	Hedges	3	19	0	1
	Picket Fence	0	0	18	5
	Wood Wall	0	0	3	5
Rejections by Class		2	1	2	6
Total Objects in Class		23	23	23	23
Errors by Class		3	3	3	12
Error Rate by Class (%)		14.29	13.64	14.29	70.59
Total Errors		21			
Average Error Rate (%)		28.20			
Total Rejections		11			

Required Class Votes Threshold = 250

		Actual Object Class			
		Brick Wall	Hedges	Picket Fence	Wood Wall
Assigned Object Class	Brick Wall	18	3	0	6
	Hedges	3	19	0	1
	Picket Fence	0	0	18	5
	Wood Wall	0	0	3	5
Rejections by Class		2	1	2	6
Total Objects in Class		23	23	23	23
Errors by Class		3	3	3	12
Error Rate by Class (%)		14.29	13.64	14.29	70.59
Total Errors		21			
Average Error Rate (%)		28.20			
Total Rejections		11			

Required Class Votes Threshold = 400

		Actual Object Class			
		Brick Wall	Hedges	Picket Fence	Wood Wall
Assigned Object Class	Brick Wall	19	3	0	6
	Hedges	2	19	0	2
	Picket Fence	0	0	19	5
	Wood Wall	0	0	3	5
Rejections by Class		2	1	1	5
Total Objects in Class		23	23	23	23
Errors by Class		2	3	3	13
Error Rate by Class (%)		9.52	13.64	13.64	72.22
Total Errors		21.00			
Average Error Rate (%)		27.25			
Total Rejections		9.00			

Table 5.5 Confusion matrices of the Adaptive Bayesian Classification Model with various threshold values for the extended objects. Fixed threshold values are noted in the upper left corner. Threshold with a varied value is noted at the upper left corner of each matrix. Thresholds highlighted in green colored text are selected as most favorable for the Adaptive Bayesian Classification Model applied to the extended objects.

(c)

Required Class Votes Threshold = 1
Ties Threshold = 0
Posterior Threshold = 0.6
Absolute Posterior Difference Threshold = 0.10

Committee Mode Frequency Threshold = 1

		Actual Object Class			
		Brick Wall	Hedges	Picket Fence	Wood Wall
Assigned Object Class	Brick Wall	17	3	0	7
	Hedges	3	19	0	1
	Picket Fence	0	0	18	5
	Wood Wall	0	0	3	5
Rejections by Class		3	1	2	5
Total Objects in Class		23	23	23	23
Errors by Class		3	3	3	13
Error Rate by Class (%)		15.00	13.64	14.29	72.22
Total Errors		22			
Average Error Rate (%)		28.79			
Total Rejections		11			

Committee Mode Frequency Threshold = 2

		Actual Object Class			
		Brick Wall	Hedges	Picket Fence	Wood Wall
Assigned Object Class	Brick Wall	17	3	0	7
	Hedges	3	19	0	1
	Picket Fence	0	0	18	5
	Wood Wall	0	0	3	5
Rejections by Class		3	1	2	5
Total Objects in Class		23	23	23	23
Errors by Class		3	3	3	13
Error Rate by Class (%)		15.00	13.64	14.29	72.22
Total Errors		22			
Average Error Rate (%)		28.79			
Total Rejections		11			

5.6 Adaptive Bayesian Classification Model Application

Committee Mode Frequency Threshold = 3

		Actual Object Class			
		Brick Wall	Hedges	Picket Fence	Wood Wall
Assigned Object Class	Brick Wall	17	3	0	7
	Hedges	3	19	0	1
	Picket Fence	0	0	18	5
	Wood Wall	0	0	3	5
Rejections by Class		3	1	2	5
Total Objects in Class		23	23	23	23
Errors by Class		3	3	3	13
Error Rate by Class (%)		15.00	13.64	14.29	72.22
Total Errors		22			
Average Error Rate (%)		28.79			
Total Rejections		11			

Committee Mode Frequency Threshold = 4

		Actual Object Class			
		Brick Wall	Hedges	Picket Fence	Wood Wall
Assigned Object Class	Brick Wall	18	3	0	6
	Hedges	3	19	0	1
	Picket Fence	0	0	18	5
	Wood Wall	0	0	3	5
Rejections by Class		2	1	2	6
Total Objects in Class		23	23	23	23
Errors by Class		3	3	3	12
Error Rate by Class (%)		14.29	13.64	14.29	70.59
Total Errors		21			
Average Error Rate (%)		28.20			
Total Rejections		11			

Committee Mode Frequency Threshold = 5

		Actual Object Class			
		Brick Wall	Hedges	Picket Fence	Wood Wall
Assigned Object Class	Brick Wall	19	3	0	6
	Hedges	2	19	0	1
	Picket Fence	0	0	19	5
	Wood Wall	0	0	3	5
Rejections by Class		2	1	1	6
Total Objects in Class		23	23	23	23
Errors by Class		2	3	3	12
Error Rate by Class (%)		9.52	13.64	13.64	70.59
Total Errors		20			
Average Error Rate (%)		26.85			
Total Rejections		10			

234   5 Adaptive Bayesian Classification Model

Table 5.5 Confusion matrices of the Adaptive Bayesian Classification Model with various threshold values for the extended objects. Fixed threshold values are noted in the upper left corner. Threshold with a varied value is noted at the upper left corner of each matrix. Thresholds highlighted in green colored text are selected as most favorable for the Adaptive Bayesian Classification Model applied to the extended objects.

(d)

Committee Mode Frequency Threshold = 5
Required Class Votes Threshold = 1
Ties Threshold = 0
Posterior Threshold = 0.6

Absolute Posterior Difference Threshold = 0.01

		Actual Object Class			
		Brick Wall	Hedges	Picket Fence	Wood Wall
Assigned Object Class	Brick Wall	19	3	0	7
	Hedges	2	20	0	1
	Picket Fence	0	0	19	5
	Wood Wall	0	0	3	6
Rejections by Class		2	0	1	4
Total Objects in Class		23	23	23	23
Errors by Class		2	3	3	13
Error Rate by Class (%)		9.52	13.04	13.64	68.42
Total Errors		21			
Average Error Rate (%)		26.16			
Total Rejections		7			

Absolute Posterior Difference Threshold = 0.10

		Actual Object Class			
		Brick Wall	Hedges	Picket Fence	Wood Wall
Assigned Object Class	Brick Wall	19	3	0	6
	Hedges	2	19	0	1
	Picket Fence	0	0	19	5
	Wood Wall	0	0	3	5
Rejections by Class		2	1	1	6
Total Objects in Class		23	23	23	23
Errors by Class		2	3	3	12
Error Rate by Class (%)		9.52	13.64	13.64	70.59
Total Errors		20			
Average Error Rate (%)		26.85			
Total Rejections		10			

5.6 Adaptive Bayesian Classification Model Application

Absolute Posterior Difference Threshold = 0.20

		Actual Object Class			
		Brick Wall	Hedges	Picket Fence	Wood Wall
Assigned Object Class	Brick Wall	19	3	0	6
	Hedges	2	19	0	1
	Picket Fence	0	0	19	5
	Wood Wall	0	0	2	4
Rejections by Class		2	1	2	7
Total Objects in Class		23	23	23	23
Errors by Class		2	3	2	12
Error Rate by Class (%)		9.52	13.64	9.52	75.00
Total Errors		19			
Average Error Rate (%)		26.92			
Total Rejections		12			

Absolute Posterior Difference Threshold = 0.30

		Actual Object Class			
		Brick Wall	Hedges	Picket Fence	Wood Wall
Assigned Object Class	Brick Wall	19	3	0	6
	Hedges	2	19	0	1
	Picket Fence	0	0	19	5
	Wood Wall	0	0	1	4
Rejections by Class		2	1	3	7
Total Objects in Class		23	23	23	23
Errors by Class		2	3	1	12
Error Rate by Class (%)		9.52	13.64	5.00	75.00
Total Errors		18			
Average Error Rate (%)		25.79			
Total Rejections		13			

Table 5.6 Confusion matrices of the Adaptive Bayesian Classification Model with various threshold values for the compact objects. Fixed threshold values are noted in the upper left corner. Threshold with a varied value is noted at the upper left corner of each matrix. Thresholds highlighted in green colored text are selected as most favorable for the Adaptive Bayesian Classification Model applied to the compact objects.

(a)

Committee Mode Frequency Threshold = 3
Required Class Votes Threshold = 250
Ties Threshold = 0
Absolute Posterior Difference Threshold = 0.10

Posterior Threshold = 0.6

		Actual Object Class	
		Steel Pole	Tree
Assigned Object Class	Steel Pole	15	0
	Tree	3	19
Rejections by Class		2	1
Total Objects in Class		20	20
Errors by Class		3	0
Error Rate by Class (%)		16.67	0
Total Errors		3	
Average Error Rate (%)		8.33	
Total Rejections		3	

Posterior Threshold = 0.7

		Actual Object Class			
		Brick Wall	Hedges	Picket Fence	Wood Wall
Assigned Object Class	Brick Wall	17	3	0	6
	Hedges	3	20	0	1
	Picket Fence	0	0	17	5
	Wood Wall	0	0	3	5
Rejections by Class		3	0	3	6
Total Objects in Class		23	23	23	23
Errors by Class		3	3	3	12
Error Rate by Class (%)		15.00	13.04	15.00	70.59
Total Errors		21			
Average Error Rate (%)		28.41			
Total Rejections		12			

5.6 Adaptive Bayesian Classification Model Application

Posterior Threshold = 0.8

		Actual Object Class	
		Steel Pole	Tree
Assigned Object Class	Steel Pole	14	0
	Tree	3	18
	Rejections by Class	3	2
	Total Objects in Class	20	20
	Errors by Class	3	0
	Error Rate by Class (%)	17.65	0
	Total Errors	3	
	Average Error Rate (%)	8.82	
	Total Rejections	5	

Posterior Threshold = 0.9

		Actual Object Class	
		Steel Pole	Tree
Assigned Object Class	Steel Pole	14	0
	Tree	2	16
	Rejections by Class	4	4
	Total Objects in Class	20	20
	Errors by Class	2	0
	Error Rate by Class (%)	12.50	0
	Total Errors	2	
	Average Error Rate (%)	6.25	
	Total Rejections	8	

238 5 Adaptive Bayesian Classification Model

Table 5.6 Confusion matrices of the Adaptive Bayesian Classification Model with various threshold values for the compact objects. Fixed threshold values are noted in the upper left corner. Threshold with a varied value is noted at the upper left corner of each matrix. Thresholds highlighted in green colored text are selected as most favorable for the Adaptive Bayesian Classification Model applied to the compact objects.

(b)

Committee Mode Frequency Threshold = 3
Ties Threshold = 0
Posterior Threshold = 0.6
Absolute Posterior Difference Threshold = 0.10

Required Class Votes Threshold = 1

		Actual Object Class	
		Steel Pole	Tree
Assigned Object Class	Steel Pole	15	0
	Tree	3	19
Rejections by Class		2	1
Total Objects in Class		20	20
Errors by Class		3	0
Error Rate by Class (%)		16.67	0
Total Errors		3	
Average Error Rate (%)		8.33	
Total Rejections		3	

Required Class Votes Threshold = 50

		Actual Object Class	
		Steel Pole	Tree
Assigned Object Class	Steel Pole	15	0
	Tree	3	19
Rejections by Class		2	1
Total Objects in Class		20	20
Errors by Class		3	0
Error Rate by Class (%)		16.67	0
Total Errors		3	
Average Error Rate (%)		8.33	
Total Rejections		3	

5.6 Adaptive Bayesian Classification Model Application

Required Class Votes Threshold = 100

		Actual Object Class	
		Steel Pole	Tree
Assigned Object Class	Steel Pole	15	0
	Tree	3	19
Rejections by Class		2	1
Total Objects in Class		20	20
Errors by Class		3	0
Error Rate by Class (%)		16.67	0
Total Errors		3	
Average Error Rate (%)		8.33	
Total Rejections		3	

Required Class Votes Threshold = 250

		Actual Object Class	
		Steel Pole	Tree
Assigned Object Class	Steel Pole	15	0
	Tree	3	19
Rejections by Class		2	1
Total Objects in Class		20	20
Errors by Class		3	0
Error Rate by Class (%)		16.67	0
Total Errors		3	
Average Error Rate (%)		8.33	
Total Rejections		3	

Required Class Votes Threshold = 320

		Actual Object Class	
		Steel Pole	Tree
Assigned Object Class	Steel Pole	15	0
	Tree	3	19
Rejections by Class		2	1
Total Objects in Class		20	20
Errors by Class		3	0
Error Rate by Class (%)		16.67	0
Total Errors		3	
Average Error Rate (%)		8.33	
Total Rejections		3	

Table 5.6 Confusion matrices of the Adaptive Bayesian Classification Model with various threshold values for the compact objects. Fixed threshold values are noted in the upper left corner. Threshold with a varied value is noted at the upper left corner of each matrix. Thresholds highlighted in green colored text are selected as most favorable for the Adaptive Bayesian Classification Model applied to the compact objects.

(c)

Required Class Votes Threshold = 1
Ties Threshold = 0
Posterior Threshold = 0.6
Absolute Posterior Difference Threshold = 0.10

Committee Mode Frequency Threshold = 1

		Actual Object Class	
		Steel Pole	Tree
Assigned Object Class	Steel Pole	15	0
	Tree	3	19
	Rejections by Class	2	1
	Total Objects in Class	20	20
	Errors by Class	3	0
	Error Rate by Class (%)	16.67	0
	Total Errors	3	
	Average Error Rate (%)	8.33	
	Total Rejections	3	

Committee Mode Frequency Threshold = 2

		Actual Object Class	
		Steel Pole	Tree
Assigned Object Class	Steel Pole	15	0
	Tree	3	19
	Rejections by Class	2	1
	Total Objects in Class	20	20
	Errors by Class	3	0
	Error Rate by Class (%)	16.67	0
	Total Errors	3	
	Average Error Rate (%)	8.33	
	Total Rejections	3	

5.6 Adaptive Bayesian Classification Model Application

Committee Mode Frequency Threshold = 3

		Actual Object Class	
		Steel Pole	Tree
Assigned Object Class	Steel Pole	15	0
	Tree	3	19
Rejections by Class		2	1
Total Objects in Class		20	20
Errors by Class		3	0
Error Rate by Class (%)		16.67	0
Total Errors		3	
Average Error Rate (%)		8.33	
Total Rejections		3	

Committee Mode Frequency Threshold = 4

		Actual Object Class	
		Steel Pole	Tree
Assigned Object Class	Steel Pole	15	0
	Tree	3	19
Rejections by Class		2	1
Total Objects in Class		20	20
Errors by Class		3	0
Error Rate by Class (%)		16.67	0
Total Errors		3	
Average Error Rate (%)		8.33	
Total Rejections		3	

Table 5.6 Confusion matrices of the Adaptive Bayesian Classification Model with various threshold values for the compact objects. Fixed threshold values are noted in the upper left corner. Threshold with a varied value is noted at the upper left corner of each matrix. Thresholds highlighted in green colored text are selected as most favorable for the Adaptive Bayesian Classification Model applied to the compact objects.

(d)

Committee Mode Frequency Threshold = 4
Required Class Votes Threshold = 1
Ties Threshold = 0
Posterior Threshold = 0.6

Absolute Posterior Difference Threshold = 0.01

		Actual Object Class	
		Steel Pole	Tree
Assigned Object Class	Steel Pole	15	0
	Tree	3	19
	Rejections by Class	2	1
	Total Objects in Class	20	20
	Errors by Class	3	0
	Error Rate by Class (%)	16.67	0
	Total Errors	3	
	Average Error Rate (%)	8.33	
	Total Rejections	3	

Absolute Posterior Difference Threshold = 0.10

		Actual Object Class	
		Steel Pole	Tree
Assigned Object Class	Steel Pole	15	0
	Tree	3	19
	Rejections by Class	2	1
	Total Objects in Class	20	20
	Errors by Class	3	0
	Error Rate by Class (%)	16.67	0
	Total Errors	3	
	Average Error Rate (%)	8.33	
	Total Rejections	3	

Absolute Posterior Difference Threshold = 0.20

		Actual Object Class	
		Steel Pole	Tree
Assigned Object Class	Steel Pole	15	0
	Tree	3	19
	Rejections by Class	2	1
	Total Objects in Class	20	20
	Errors by Class	3	0
	Error Rate by Class (%)	16.67	0
	Total Errors	3	
	Average Error Rate (%)	8.33	
	Total Rejections	3	

5.6 Adaptive Bayesian Classification Model Application 243

Absolute Posterior Difference Threshold = 0.30

		Actual Object Class	
		Steel Pole	Tree
Assigned Object Class	Steel Pole	15	0
	Tree	3	19
Rejections by Class		2	1
Total Objects in Class		20	20
Errors by Class		3	0
Error Rate by Class (%)		16.67	0
Total Errors		3	
Average Error Rate (%)		8.33	
Total Rejections		3	

The thresholds provide the ability to fine tune the classification model to support the extended and compact object categories. The appropriate selection of the threshold values will minimize the classification error rate and number of rejections. For the extended objects, we fixed the threshold values as displayed in the upper left corner of Table 5.5 (a) and varied the Posterior Threshold as shown. As we see, a Posterior Threshold = 0.6 (or 60%) provides the most favorable average error rate (28.20%) and the least amount of rejections of class assignments (11). We now fix Posterior Threshold = 0.6 and vary the Required Class Votes Threshold as shown in Table 5.5 (b). In this case, the Required Class Votes Threshold set to 400 results in the lowest average error rate and total rejections. However, we will subjectively select Required Class Votes Threshold =1 since the setting of 400 appears to slightly increase the number of misclassifications for the wood wall object class, which is already more vulnerable to classification errors as discussed in Sect. 5.4. Fixing the Required Class Votes Threshold = 1, we now vary the Committee Mode Frequency Threshold as displayed in Table 5.5 (c). As we see, setting Committee Mode Frequency Threshold = 5 provides the most favorable performance. Table 5.5 (d) presents variations of our final threshold, Absolute Posterior Difference Threshold, while fixing the threshold displayed in the upper left corner of the matrices. We see that the threshold settings in the upper left corner along with letting Absolute Posterior Difference Threshold = 0.01 provide an acceptable average error rate of 26.16% with only a total of 7 rejections of class assignments. Consequently, these threshold settings appear to be a favorable selection for our extended objects.

Analysis of the performance of the Adaptive Bayesian Classification Model on the compact objects with variations in the threshold values was conducted in a similar fashion as the extended objects. For the compact objects, we begin in Table 5.6 (a) by fixing the thresholds displayed in the upper left corner and varying the Posterior Threshold. As we see, setting the Posterior Threshold = 0.6 (or 60%) results in the most favorable average error rate and total rejections. Consequently, we will choose 0.6 as the setting for the Posterior Threshold. Since the variations in the Required Class Votes Threshold and Committee Mode Frequency Threshold in Table 5.6 (b) and 5.6 (c), respectively, do not show any changes in the classification performance, we will set each of the their thresholds equal to one. In Table 5.6 (d), we also see that the variations in the Absolute Posterior Difference Threshold values do not produce any changes in the model's classification performance. Consequently, we will choose the settings of the thresholds in the upper left corner and Absolute Posterior Difference Threshold = 0.01 as our favorable choices for the compact objects.

Tables 5.7 and 5.8 provide a comparison of the confusion matrices of our Adaptive Bayesian Classification Models with the threshold settings discussed above to the best performers among the Adaptive Bayesian Classifier with the single distance function, KNN Classifier, and Parzen Classifier from Tables 5.1 and 5.2 on the extended and compact object categories, respectively. As we see, our Adaptive Bayesian Classification Model performs exceptionally on the blind extended and compact objects shown in Table 2.2 compared to the KNN Classifier and Parzen Classifier. While the committees of experts and dynamical window technique integrated into the Adaptive Bayesian Classification Model increase the accuracy of class assignments and our confidence in the model's final classification decision, the ability to reject class assignments that do not satisfy specific rules is the distinguishing factor that results in the Adaptive Bayesian Classification Model outperforming the Adaptive Bayesian Classifier with a single distance function.

Table 5.7 Comparison of confusion matrices of the best performing classification models applied to the extended objects from the Adaptive Bayesian Classification Model (via Committees of Experts), Adaptive Bayesian Classifier with single distance function, KNN Classifier, and Parzen Classifier.

(a)

Adaptive Bayesian Classification Model
Committee Mode Frequency Threshold = 5
Required Class Votes Threshold = 1
Ties Threshold = 0
Posterior Threshold = 0.6
Absolute Posterior Difference Threshold = 0.01

		Actual Object Class			
		Brick Wall	Hedges	Picket Fence	Wood Wall
Assigned Object Class	Brick Wall	19	3	0	7
	Hedges	2	20	0	1
	Picket Fence	0	0	19	5
	Wood Wall	0	0	3	6
Rejections by Class		2	0	1	4
Total Objects in Class		23	23	23	23
Errors by Class		2	3	3	13
Error Rate by Class (%)		9.52	13.04	13.64	68.42
Total Errors		21			
Average Error Rate (%)		26.16			
Total Rejections		7			

5.6 Adaptive Bayesian Classification Model Application 245

(b)

Adaptive Bayesian Classifier
<1,2,3,4,6,11,13,14,16>, D_I

		Actual Object Class			
		Brick Wall	Hedges	Picket Fence	Wood Wall
Assigned Object Class	Brick Wall	18	3	0	8
	Hedges	5	20	1	3
	Picket Fence	0	0	22	6
	Wood Wall	0	0	0	6
Total Objects in Class		23	23	23	23
Errors by Class		5	3	1	17
Error Rate by Class (%)		21.74	13.04	4.35	73.91
Total Errors		26			
Average Error Rate (%)		28.26			

(c)

KNN Classifier
<3,5,6,7,8,9,13,16>, $K = 3$

		Actual Object Class			
		Brick Wall	Hedges	Picket Fence	Wood Wall
Assigned Object Class	Brick Wall	15	3	0	5
	Hedges	6	20	0	8
	Picket Fence	0	0	20	4
	Wood Wall	2	0	3	6
Total Objects in Class		23	23	23	23
Errors by Class		8	3	3	17
Error Rate by Class (%)		34.78	13.04	13.04	73.91
Total Errors		31			
Average Error Rate (%)		33.70			

(d)

Parzen Classifier
<1,3,4,6,11,12,14,18>, $h = 0.044$

		Actual Object Class			
		Brick Wall	Hedges	Picket Fence	Wood Wall
Assigned Object Class	Brick Wall	14	3	0	5
	Hedges	8	18	1	5
	Picket Fence	0	1	21	5
	Wood Wall	1	1	1	8
Total Objects in Class		23	23	23	23
Errors by Class		9	5	2	15
Error Rate by Class (%)		39.13	21.74	8.70	65.22
Total Errors		31			
Average Error Rate (%)		33.70			

Table 5.8 Comparison of confusion matrices of the best performing classification models applied to the compact objects from the Adaptive Bayesian Classification Model (via Committees of Experts), Adaptive Bayesian Classifier with single distance function, KNN Classifier, and Parzen Classifier.

(a)

Adaptive Bayesian Classification Model			
Committee Mode Frequency Threshold = 4			
Required Class Votes Threshold = 1			
Ties Threshold = 0			
Posterior Threshold = 0.6			
Absolute Posterior Difference Threshold = 0.01			
		Actual Object Class	
		Steel Pole	Tree
Assigned Object Class	Steel Pole	15	0
	Tree	3	19
	Rejections by Class	2	1
	Total Objects in Class	20	20
	Errors by Class	3	0
	Error Rate by Class (%)	16.67	0
	Total Errors	3	
	Average Error Rate (%)	8.33	
	Total Rejections	3	

(b)

Adaptive Bayesian Classifier			
<1,2,4,6,7,8,10,11,13,14,15>, D_l			
		Actual Object Class	
		Steel Pole	Tree
Assigned Object Class	Steel Pole	16	0
	Tree	4	20
	Total Objects in Class	20	20
	Errors by Class	4	0
	Error Rate by Class (%)	20.00	0
	Total Errors	4	
	Average Error Rate (%)	10.00	

(c)

KNN Classifier			
<6,7,12,14>, $K = 7$			
		Actual Object Class	
		Steel Pole	Tree
Assigned Object Class	Steel Pole	14	0
	Tree	6	20
	Total Objects in Class	20	20
	Errors by Class	6	0
	Error Rate by Class (%)	30.00	0
	Total Errors	6	
	Average Error Rate (%)	15.00	

(d)

Parzen Classifer			
<6,7,9,14>, $h = 0.0379$			
		\multicolumn{2}{c}{Actual Object Class}
		Steel Pole	Tree
Assigned Object Class	Steel Pole	14	0
	Tree	6	20
	Total Objects in Class	20	20
	Errors by Class	6	0
	Error Rate by Class (%)	30.00	0
	Total Errors	6	
	Average Error Rate (%)	15.00	

5.6.2 Performance on Blind Data (with Classes ≠ Training Set)

We will now evaluate the Adaptive Bayesian Classification Model's response when confronted with the following additional blind objects that include objects outside the classes in the training data sets. Figure 5.15 displays the visible and thermal images of some random blind extended objects consisting of a brick wall with moss on the surface, concrete wall, bush, gravel pile, steel picket fence, wood bench, and wood wall of a storage shed. Figure 5.16 displays the visible and thermal images of some blind compact objects consisting of a square steel pole, aluminum pole for a dryer vent, concrete pole, knotty tree, telephone pole, 4×4 wood pole, and pumpkin. The thermal images of these objects were captured between 6 July and 5 November 2007 on The College of William & Mary campus, throughout York County, Virginia, and on a farm outside Buffalo, New York. The performance of the Adaptive Bayesian Classification Model was assessed on these objects with threshold settings selected as discussed above for the extended and compact object categories and displayed in the confusion matrices for the Adaptive Bayesian Classification Model in Tables 5.7 and 5.8.

Table 5.9 presents the actual blind object and object class assigned by the Adaptive Bayesian Classification Model along with the resulting posterior probability for the extended objects. The brick wall with moss on the surface in Fig. 5.15 (a) was misclassified as hedges due to the high thermal radiance of the reference emitter and the high thermal radiance and contrast associated with the blind brick wall having a strong resemblance to those of the hedges in the model's training data set as discussed in Sect. 5.4. The classification performance on brick walls with a larger range of thermal radiances could be improved by increasing the range of representative objects in the training data set as noted in Sect. 5.4.2.3. Furthermore, since the posterior probability for assigning the brick wall as hedges was 81.78%, setting the model's Posterior Threshold to 82% would result in the model rejecting this class assignment and requiring the bot to capture another thermal image of the brick wall for classification, perhaps at another viewing angle. The concrete wall in Fig. 5.15 (b) and bush in Fig. 5.15 (c) were appropriately classified as a brick wall and hedges, respectively. The gravel pile in Fig. 5.15 (d) classified as hedges

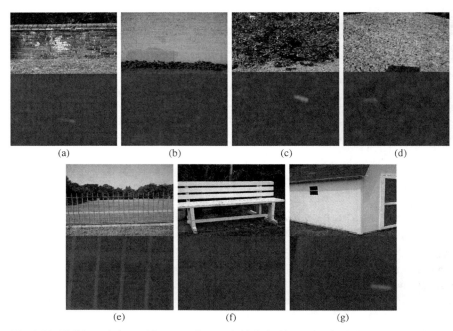

Fig. 5.15 Visible and thermal images of extended blind objects that include classes outside the given training data set. (a) brick wall with moss on the surface, (b) concrete wall, (c) bush, (d) gravel pile, (e) steel picket fence, (f) wood bench, and (g) wood wall of a storage shed.

since it displays variations (randomness or complexity) in radiant emissions that are similar to the hedges in the training data set. Even though the thermal-physical properties of the steel picket fence used for the blind data in Fig. 5.15 (e) and wood picket fence used in our training data set are obviously different, the model appropriately classified the steel picket fence as a (wood) picket fence since the blind object has the same picket pattern and similar thermal emissions from the foreground as the wood picket fences in the training data.

During our research, we have continuously emphasized our desire to design a classification model that affords the ability to retain the original physical interpretation of the information in the signal data throughout the entire classification process. As a result, our Adaptive Bayesian Classification Model provides the ability to analyze the physical characteristics of objects and decisions by the experts in each committee to understand the reason for misclassifications and rejections of class assignments. As we see in Table 5.9, the Adaptive Bayesian Classification Model rejected the class assignment for the wood bench in Fig. 5.15 (f) and wood wall on the shed in Fig. 5.15 (g). By analyzing each committee's decision making process, we found that the classifications of both the wood bench and wood wall were rejected for not satisfying specific rules within each committee. The class assignment of the wood bench was rejected since each committee did not have a mode class equal to their respective object class. For instance, the mode

Fig. 5.16 Visible and thermal images of compact blind objects that include classes outside the given training data set. (a) square steel pole, (b) aluminum pole for dryer vent, (c) concrete pole, (d) knotty tree, (e) telephone pole, (f) 4×4 wood pole, and (g) pumpkin.

class for the brick wall, hedges, and picket fence committees was the wood wall. The mode class for the wood wall committee was the picket fence. As a result, the mode class rule was not satisfied and each committee recommended a class label of 0 to the wood bench with a posterior probability of 0%. The model subsequently rejected the classification of the wood bench. Similarly, the classification of the wood wall was also rejected for not satisfying rules within each committee. In this case, the brick wall, picket fence, and wood wall committees each recommended a class label of 0 and posterior probability of 0% to the wood wall since each of their mode classes was the hedges. The hedges committee had a mode class of hedges; however, a class label of 0 and posterior probability of 0% was recommended since only four out of the required five experts had the hedges as their mode class. Consequently, the model rejected the classification of the wood wall.

Table 5.10 presents the actual blind object and object class assigned by the Adaptive Bayesian Classification Model along with the resulting posterior probability for the compact objects. As we see, the square steel pole in Fig. 5.16 (a) was appropriately classified as a steel pole by the model. We would expect an aluminum pole to classify as a steel pole due to its approximately constant thermal radiance on the surface. However, as we see in Fig. 5.16 (b), the expected constant thermal radiance on the surface is interrupted by a crease in the aluminum that results in a higher thermal radiance emitted from the crease due to the variation of emissivity

Table 5.9 (a) Adaptive Bayesian Classification Model class assignments and posterior probabilities on extended blind objects displayed in Fig. 5.15. (b) Threshold values for the Adaptive Bayesian Classification Model.

Actual Object Class	Assigned Object Class	Posterior Probability (%)
Brick Wall w/ Moss	Hedges	81.78
Concrete Wall	Brick Wall	96.23
Bush	Hedges	99.38
Gravel Pile	Hedges	92.00
Steel Picket Fence	Picket Fence	97.61
Wood Bench	(Class Assignment Rejected)	0
Wood Wall on Shed	(Class Assignment Rejected)	0

(a)

Threshold	Value
Committee Mode Frequency	5
Required Class Votes	1
Ties	0
Posterior	60%
Absolute Posterior Difference	0.01

(b)

Table 5.10 (a) Adaptive Bayesian Classification Model class assignments and posterior probabilities on compact blind objects displayed in Fig. 5.16. (b) Threshold values for the Adaptive Bayesian Classification Model.

Actual Object Class	Assigned Object Class	Posterior Probability (%)
Square Steel Pole	Steel Pole	99.99
Aluminum Pole	Tree	96.67
Concrete Pole	Steel Pole	99.65
Knotty Tree	Tree	99.58
Telephone Pole	(Class Assignment Rejected)	0
4x4 Wood Pole	Tree	85.54
Pumpkin	Steel Pole	77.88

(a)

Threshold	Value
Committee Mode Frequency	4
Required Class Votes	1
Ties	0
Posterior	60%
Absolute Posterior Difference	0.01

(b)

with the shape of the object as we discussed in Chap. 3. Consequently, the model sees the thermal features generated from the surface of the aluminum pole more closely resembling the features of the trees in the training data set. The concrete pole in Fig. 5.16 (c) classified as a steel pole due to its approximately constant thermal radiance on the surface resembling the surfaces of the steel poles in the training data. The knotty tree in Fig. 5.16 (d) classified as a tree as expected. As we see in Table 5.10, the model rejected the classification of the telephone pole in Fig. 5.16 (e). By analyzing the execution of the Adaptive Bayesian Classification Model on the telephone pole, we learned that both the steel pole and tree committees had a mode class equal to the steel pole. As a result, the tree committee recommended a class label of 0 and posterior of 0% to the telephone pole. On the other hand, the rules for the mode class, committee mode frequency threshold, required class vote threshold, and ties threshold were satisfied within the steel pole committee. However, the steel pole committee's mean posterior probability for the telephone pole was only 53.76%. Therefore, the rule with the Posterior Threshold set to 60% was not satisfied and the steel pole committee also recommended a class label of 0 and posterior of 0% to the telephone pole. The final decision by the Adaptive Bayesian Classification Model was to reject the classification of the telephone pole. The 4×4 wood pole in Fig. 5.16 (f) classified as a tree by the model as expected. The pumpkin in Fig. 5.16 (f) classified as a steel pole since the model saw the pumpkin's surface, with an approximately constant thermal radiance, resembling the thermal radiance and contrast typically found on the surfaces of the steel poles in the training data set as displayed in Fig. 5.11 (a–c). Fortunately, our Adaptive Bayesian Classification Model is equipped with the rule involving the Posterior Threshold. As a result, a simple tuning that sets the model's Posterior Threshold to say 80% will let the bot reject the classification of the pumpkin.

5.7 Summary

The concepts, methods, and thermal features introduced in the previous chapters culminated in the design and implementation of the novel pattern classification tools presented in this chapter that can be used to understand the behavior of the thermal patterns of non-heat generating object classes in an n-dimensional feature space and classify an unknown pattern that is mapped into the feature space. In this chapter, we first showed how to apply principal component analysis locally on the patterns from a given object class to derive two distance metrics – based on a scalar projection (Eq. 5.1) and normal distance (Eq. 5.2) involving the patterns and first principal eigenvectors in feature space. We showed how these distance metrics provide the ability to see and understand the behavior of an object class's patterns about its first principal eigenvector that projects through the respective hyperconoidal cluster. Additionally, we demonstrated how our distance metrics give us the ability to "see" regions in an n-dimensional feature space where some object classes may tend to "look alike" and run the risk for misclassification by a classification model.

Various distance functions $d_j(\widetilde{f}, \underline{e}_{1j})$ were derived based on the normal distance between patterns and an object class's first principal eigenvector. These distance functions were incorporated into the likelihood function of the Bayesian classifiers to form our adaptive Bayesian classifier given by Eq. 5.10. In this way, we formed a weighted likelihood function used in the posterior probability of the Bayesian classifier that not only considers the unknown pattern's participation in the density distribution of a given object class but also the unknown pattern's behavior about the first principal eigenvector projecting through the given object class's hyperconoidal cluster. The variations of the distance functions were designed to adapt to the behavior of the patterns for a given object class, as the name for the adaptive Bayesian classifier implies. The resulting adaptive Bayesian classifier with the weighted likelihood function was shown to produce a posterior probability with enhanced discriminating capabilities that outperformed the traditional KNN and Parzen classifiers.

As we have stated in previous chapters, the performance of a classifier is a function of the feature vector. However, rather than analyzing the classification performance by just choosing different feature vectors, the novel process used by our adaptive Bayesian classifier affords us the ability to literally *see* how the choice of any n-dimensional feature vector will affect the behavior of an object class's patterns and the overall performance of the classification model. As we discussed in Sect. 5.2, the distance metrics, given by Eqs. 5.1 and 5.2, give us the ability to see a general trend in the behavior of the patterns within each object class that vary slightly depending on the feature vector. Thus, the behavior of the patterns about the first principal eigenvector is dependent on the choice of the n-dimensional feature vector. Consequently, the normal distance metric, given by Eq. 5.2, depends on the behavior of the patterns about the first principal eigenvector. The normal distance metric has an effect on the values of our distance function $d_j(\tilde{f}, \underline{e}_{1j})$ and the weighted KNN density estimation given by Eq. 5.9. As a result, the classification performance of our model is based on the values of our adaptive Bayesian classifier, given by Eq. 5.10, that are dependent on the weighted KNN density estimation. Therefore, the performance of a classifier is a function of the feature vector.

We used our distance metrics and adaptive Bayesian classifier to understand why some blind patterns are being misclassified under certain thermal conditions. We noted that correct classification of a blind object seemed to be independent of the geographical location of the object. Thus, the two primary factors that contributed to the misclassification of the blind objects were a lack of representative training data and the effects of the diurnal cycle of solar energy. Consequently some misclassifications could be eliminated by expanding the range of features in the training data set by capturing a more representative set of thermal images. However, in most cases a lack of a thermal signature from an object due to the diurnal cycle of solar energy will continue to result in feature values from different object classes looking alike. We also observed that in some cases the misclassification of a blind object was associated with either a low posterior probability or a posterior that was close in value to another posterior for an assignment to a different object class. These situations led to our integration of specific rules into our novel classification model and our plans for future research involving the integration of fuzzy logic into our model and designing a model based on a multi-sensor data fusion architecture that we will discuss in Chap. 6.

Based on our discovery that some adaptive Bayesian classifiers act as experts by showing exceptional classification performance on a certain object class, we formed committees of experts where each committee classifies patterns from their respective object class. By combining each committee of experts into one classification model, we were able to exploit the expertise of each committee and complement the overall performance of the classification model. We further increased the confidence level in our model's classification decisions by integrating the dynamical window technique presented in Chap. 4 that lets each committee of experts decide on class assignment by considering information collected from multiple window sizes of the thermal image of an object. Additionally, we incorporated rules into our model that must be satisfied before the bot is authorized to make a classification decision to improve the accuracy of class assignments and prevent high-risk classification decisions. If all the rules are satisfied, the bot is authorized to assign a class to the unknown object within its field of view and proceed with the next required action in the intelligence algorithm. On the other hand, if a rule is not satisfied, the bot must reject the class assignment and capture another thermal image of the unknown object for classification, perhaps at another viewing angle. These concepts led to the design of our novel *Adaptive Bayesian Classification Model* displayed in Fig. 5.14.

By assessing our Adaptive Bayesian Classification Model on extended and compact blind data that consisted of objects from the same and different object classes as the training data, we proved the exceptional applicability and originality of our model. Our application demonstrated that the Adaptive Bayesian Classification Model outperforms the traditional KNN Classifier and Parzen Classifier. Additionally, while the committees of experts and dynamical window technique integrated into the Adaptive Bayesian Classification Model increase the accuracy of class assignments and our confidence in the model's final classification decision, the ability to reject class assignments that do not satisfy specific rules is the distinguishing factor that results in the Adaptive Bayesian Classification Model outperforming the Adaptive Bayesian Classifier with a single distance function.

The design of our Adaptive Bayesian Classification Model makes it an appropriate method to support multiple scenarios. First, the Adaptive Bayesian Classification Model is a suitable choice for any classification application, such as ours, involving hyperconoidal clusters consisting of patterns in an n-dimensional feature space that are characterized by their behavior about their respective first principal eigenvector. Such applications involve features that vary due to the effects of some natural cyclic events. The natural cyclic event in our application is the diurnal cycle of solar energy. Furthermore, the emphasis on designing the model so that the original physical interpretation of the information in the signal data is retained throughout the entire classification process affords human operators the ability to analyze the reason for a bot's class assignments by associating the final classification decision with the thermal-physical properties found in the original features. Also, the integration of the dynamical window technique and classification rules with the option to reject class assignments and capture another thermal image of the unknown object for classification, perhaps at another viewing angle, make our model appropriate for autonomous robotic systems that capture continuous frames.

The design and implementation of our Adaptive Bayesian Classification Model has also created new research opportunities. Research is required to determine if there exists a most favorable number of experts in each committee. Also, the selection of the most favorable threshold values requires additional research. The appropriate selection of threshold values will minimize the classification error rate and number of rejections. The tendency for an object to "look like" another object under certain thermal conditions (other than thermal crossover) presents a degree of vagueness that may call for the integration of fuzzy logic into the classification model. We could also integrate other sensors into the autonomous robotic system by designing a multi-sensor data fusion architecture where the use of multiple sensors complements the overall performance of the classification model. We will discuss these research opportunities in our final chapter, Chap. 6.

References

[1] Gavert H, Hurri J et al (2005) FastICA for Matlab 7.x and 6.x Version 2.5. Helsinki University of Technology, Finland
[2] Duin RPW, Juszczak P et al (2004) PRTools4, A Matlab Toolbox for Pattern Recognition. Delft University of Technology, The Netherlands
[3] Holst GC (2000) Common Sense Approach to Thermal Imaging. JCD Pub.; co-published by SPIE Optical Engineering Press, Winter Park, Fla.; Bellingham, Wash.
[4] Loftsgaarden DO, Quesenberry CP (1965) A Nonparametric Estimate of a Multivariate Density Function. The Annals of Mathematical Statistics 36(3):1049–1051
[5] Theodoridis S, Koutroumbas K (2006) Pattern Recognition. 3rd edn. Academic Press, San Diego, CA
[6] Webb AR (2002) Statistical Pattern Recognition. 2nd edn. Wiley, West Sussex, England; New Jersey
[7] Duda RO, Hart PE et al (2001) Pattern Classification. 2nd edn. Wiley, New York

6 Conclusions and Future Research Directions

Abstract This book presented the design and implementation of a physics-based adaptive Bayesian pattern classification model that uses a passive thermal infrared imaging system to automatically characterize non-heat generating objects in unstructured outdoor environments for mobile robots. The resulting model complements an autonomous robot's situational awareness and affords bots with the intelligence to automatically interpret the information in signal data emitted from targets to make decisions without the need for an interpretation by humans. The work presented in this book has created new opportunities to continue the research in support of the goal to automate the fusion and interpretation of data streams from various active and passive sensor systems to enable autonomous mobile robot operations in a wide variety of unstructured outdoor environments.

6.1 Introduction

In this book, we have designed and implemented a novel pattern classification model to characterize non-heat generating outdoor objects in thermal scenes for application to autonomous robots. In the context of this research, we have defined non-heat generating objects as objects that are not a source for their own emission of thermal energy, and so exclude people, animals, vehicles, etc. The resulting model complements the autonomous bot's situational awareness that supports decision-making in the overall intelligence process. In this final chapter, we will summarize the research contributions of this work, identify the primary limitation to using a thermal infrared imaging system in our application, and discuss our future research directions.

6.2 Contributions

We have developed a set of methods and algorithms that use a thermal infrared imaging system to automatically characterize non-heat generating extended and compact objects in outdoor environments. The extended objects consisted of objects that extend laterally beyond the thermal camera's lateral field of view, such as brick walls, hedges, picket fences, and wood walls. The compact objects consisted of objects that are completely within the thermal camera's lateral field of view, such as steel poles and trees. We included a systematic and detailed analysis on the acquisition and preprocessing of thermal images, generation and selection of thermal-physical features from these non-heat generating objects within thermal images, and the design of a novel physics-based model to automatically classify these objects. Many of our concepts and methods evolved by integrating techniques from various fields of study, such as thermography and pattern classification, to gain an understanding of the underlying physical behavior of the information in the thermal signal produced by a non-heat generating object. During our research, we also designed our classification model to retain the original physical interpretation of the information in the signal data throughout the entire classification process. This emphasis resulted in a framework that allows the analyst to understand the reason for a bot's classification of an unknown object by associating the final classification decision with the thermal-physical properties found in the original features. Additionally, our approach affords bots with the intelligence to automatically interpret the information in signal data to make decisions without rendering high-quality imagery for human experts to interpret.

Three primary contributions from this research are: (1) an Adaptive Bayesian Classification Model, (2) distance metrics used to describe the behavior of an object class's patterns about the eigenvector that projects through its respective hyperconoidal cluster, and (3) a curvature algorithm that will allow us to distinguish compact objects from extended objects. Our Adaptive Bayesian Classification Model presented in Chap. 5 outperformed the traditional KNN and Parzen classifiers. The design of our Adaptive Bayesian Classification Model makes it an appropriate method to support multiple scenarios. First, the Adaptive Bayesian Classification Model is a suitable choice for any classification application, such as ours, involving hyperconoidal clusters consisting of patterns in an n-dimensional feature space that are characterized by their behavior about their respective first principal eigenvector. Such applications involve features that vary due to the effects of some natural cyclic events. Our model is designed to adapt to the behavior of these patterns from specified object classes to provide an accurate classification of unknown objects. Furthermore, the emphasis on designing the model so that the original physical interpretation of the information in the signal data is retained throughout the entire classification process affords human operators the ability to analyze the reason for a bot's class assignments by associating the final classification decision with the thermal-physical properties found in the original features. Also, the integration of the dynamical window technique and classification rules with the option to reject class assignments and capture another thermal image of the unknown object for classification, perhaps at another viewing angle, make our model appropriate for autonomous robotic systems that capture continuous frames.

The two distance metrics, based on the scalar projection (Eq. 5.1) and normal distance (Eq. 5.2), were a precursor to our Adaptive Bayesian Classification Model. These two distance metrics give us the ability to "see" and understand the behavior of an object class's patterns within their respective hyperconoidal cluster in an n-dimensional feature space. Additionally, we demonstrated how our distance metrics give us the ability to "see" regions in an n-dimensional feature space where some object classes may tend to "look alike" and run the risk for misclassification by a classification model. Consequently, these metrics provide the researcher with a technique to analyze and select n-dimensional feature vectors as well as predict the classification performance of a given model when using the selected feature vectors.

In Chap. 3, we introduced a curvature algorithm that allows us to distinguish compact objects from extended objects. During our analysis involving the generation of thermal features used by our classification model, we discovered that certain factors caused variations in radiance on cylindrical-shaped objects. These factors, consisting of directional variation of emissivity, irradiance from sources in the background, and/or halo effect, assisted us in deriving a curvature algorithm used to distinguish compact objects from extended objects. In the context of this research, we defined background as the region either in front or to the side of the target consisting of thermal sources that emit thermal energy onto the target's surface. The source emitting this thermal energy may or may not be in the camera's field of view. On the other hand, we defined foreground as the region in the scene consisting of objects behind the target of interest and within the thermal camera's field of view. Our curvature algorithm is presented in Table 3.5. A demonstration of the curvature algorithm showed that we were able to correctly identify a tree and square metal pole as compact objects and a brick wall as an extended object. With further investigation the curvature algorithm has potential to serve as an exceptional technique to distinguish compact objects from extended objects.

6.3 Limitation of a Thermal Infrared Imaging System

Understanding the limitations of sensor systems used by any pattern classification model is important since depending on the environmental conditions the sensor may not be able to obtain relevant features to classify an unknown object due to the lack of signal information emitted from the object. In this case, our autonomous robot may have to rely on its other sensor(s) to classify the object. Since our application takes place outdoors, environmental conditions will exist where the surfaces of a target and surrounding objects will emit approximately the same level of thermal radiance. This phenomenon, known as thermal crossover [1], results in minimal thermal contrast between the surfaces of objects and the surrounding environment within the thermal infrared camera's field of view. Thermal images of objects captured during thermal crossover run the risk of producing features that the bot will attribute to features from other object classes. Thermal crossover was a factor that contributed to the misclassification of our blind objects in Chap. 5 and is seen as the primary limitation in our ability to accurately classify non-heat generating objects in an outdoor environment using a thermal imaging system.

Thermal crossover will always occur as part of the natural diurnal cycle of solar energy. The length of time that the phenomenon occurs depends on the thermal properties of objects' surfaces, time history of solar radiation, and time of day. Environmental conditions such as low ambient temperatures and/or lack of direct solar energy on an object's surface (e.g., due to shady locations, clouds, or night time) reduce an object's emission of thermal radiance. Our ability to detect objects in thermal images captured at night depends on the thermal properties of the object and the time history of solar radiation. Thus, as we discussed in Chap. 3, the amount of thermal radiance emitted by an object depends on the emissivity of the object. The higher an object's emissivity, the more thermal radiance it will emit. Emissivity depends on surface temperature (as well as the type of material, viewing angle, and the object's surface quality and shape) and surface temperature depends on the specific heat (as well as conductivity and other thermal properties) of the object. Objects with a high specific heat, such as birch trees ($\sim 2.4\ kJ \cdot kg^{-1} \cdot {}^\circ C^{-1}$) [2], will tend to heat up more slowly with the increasing solar energy and cool more slowly as the amount of solar energy begins to decrease in the late afternoon (around 1600 hrs). On the other hand, the surface temperature of low specific heat objects, such as the leaves on hedges, tend to track the availability of solar energy [1]. When a cloud passes or the sun begins to set, the surface temperature of the hedges stays consistent with the lower ambient temperature. As a result, a low level of solar energy available to a low specific heat object results in less thermal radiation emitted. If a birch tree and hedges exist side-by-side and are in direct sunlight in the afternoon on a summer's day, an acceptable thermal contrast will exist in the scene to detect, segment, and classify both objects. Since the birch tree will emit more thermal radiance than the hedges after sunset, there will still exist enough thermal contrast between the two objects in the scene to segment the birch tree. However, the bot will more likely only be able to generate relevant thermal features from the surface of the birch tree. On a cloudy day with a low ambient temperature in the winter, both the birch tree and hedges will emit minimal thermal radiation. In this case, there will likely not exist enough thermal contrast in the scene for the bot to distinguish the two objects. An attempt to classify the objects in the scene will thus result in misclassifications.

The best way to deal with periods of thermal crossover is have the bot avoid using the thermal infrared imaging modality when minimal thermal contrast exists in the scene. A feasible course of action would be to integrate a thermal contrast threshold rule into the detection phase of the intelligence process that requires a minimum amount of contrast in the scene to use the thermal infrared imaging modality. If the rule is not satisfied, the bot must eliminate the use of the thermal infrared imaging sensor and rely on other sensors, such as ultrasound, that are available in the multi-sensor data fusion framework to classify this specific target. The limitations found with any sensor obviously provide the reason why multi-sensor data fusion systems are normally more successful in classification applications than systems with a single sensor. Thus, the interpretations of relevant information received by different types of sensors used in a multi-sensor framework are fused to complement the overall performance of the classification process. In Sect. 6.4, we will discuss our plans for integrating our current pattern classification model using thermal infrared imagery into a multi-senor data fusion framework.

6.4 Future Research

The work presented in this book has created new opportunities to continue the research in support of the goal to automate the fusion and interpretation of data streams from various active and passive sensor systems to enable autonomous mobile robot operations in a wide variety of unstructured outdoor environments as discussed in Chap. 1. In this section, we will discuss our future research directions that evolve from our current work and research involving sonar sensor interpretation by mobile robots [3].

6.4.1 Augmentation of Robotic Thermal Imaging System

The design and implementation of our Adaptive Bayesian Classification Model has created new research opportunities. Research is required to determine if there exists a most favorable number of experts in each committee. Also, the selection of the most favorable threshold values requires additional research. The appropriate selection of threshold values will minimize the classification error rate and number of rejections.

Although our current research involved a parked robot capturing still frames and then moving to the next location before capturing another still frame, it is not difficult to envision a similar mobile robotic system that interprets objects in thermal images captured from continuous frames while moving. The robotic system could then capture thermal images at a frame rate of 30 images per second. Continuous frames would afford the bot with a "real-time" classification and quick response to capture another thermal image of an object that was previously rejected by the Adaptive Bayesian Classification Model for not satisfying the rules for a class assignment.

Research involving classifying unknown objects from continuous frame will require the integration of detection and segmentation algorithms into the algorithm of the classification model. In this work, we assumed that the bot had already detected and segmented an unknown object. There are many options for integrating detection and segmentation algorithms into the overall classification framework. In Chap. 1, we presented references that discuss detection and segmentation methods using various passive and active modalities, such as thermal infrared, RGB, and sonar sensors. In Chap. 2, we discussed how the halo effect, resulting from the mechanical chopper wheel within a thermal infrared camera, could produce a halo around targets. Consequently, this halo effect could serve to assist the bot in segmenting a target for classification [4]. Additionally, we will also automate the classification process to detect, segment, and classify targets in cluttered scenes.

As we discussed in Chap. 5, the two primary factors that contributed to the misclassification of the blind objects were a lack of representative training data and the effects of thermal crossover. The integration of a thermal crossover threshold rule to avoid misclassifications due to thermal crossover was introduced in Sect. 6.3. Thus, future research involving the use of thermal infrared imaging system will also need to include an expanded range of features in the training data set by capturing a more representative set of thermal images.

The current robotic thermal imaging system design uses electrical tape as a reference emitter and crinkled aluminum foil to estimate the irradiance received by the target. The electrical tape and crinkled aluminum foil are attached to the target to capture their thermal images used to generate the required feature values discussed in Chap. 3. Research is required to determine how to estimate the thermal radiance emitted from a reference emitter and capture the irradiance received by the target without the need to pre-attach the electrical tape and aluminum foil.

6.4.2 Fuzzy Logic Classifier

Research is required to explore the integration of a fuzzy logic classifier into the Adaptive Bayesian Classification Model. This research would be based on the observations, in Chaps. 4 and 5, that the classification models consistently misclassified some patterns from specific object classes while other patterns were assigned to the correct class. We have determined that some object classes look alike when operating "beyond the visible spectrum" under certain thermal conditions (other than thermal crossover). These conditions result in objects that are imprecisely defined. For instance, under certain thermal conditions the feature vectors from a wood wall may look like a brick wall, and a picket fence under other conditions. This type of uncertainty presents a degree of vagueness that may call for the integration of fuzzy logic into the classification model [5, 6].

We could introduce our use of fuzzy logic and membership functions based on a feature called *sparsity* that is generated from the 2-dimensional frequency spectrum of an object's thermal image [3]. Four sparsity features can be generated from an object's thermal image to measure how well defined the edge directions are on the object. After pre-processing the object's thermal image as discussed in Chap. 2, we take

Fig. 6.1 (a) visible image, (b) thermal images, (c) frequency spectrum, and (d) polar spectrum of a wood wall.

the 2D Fourier transform of the object's thermal image and take the absolute value to obtain the spectrum, which is then transformed to polar coordinates with angle measured in a clockwise direction from the polar axis and increasing along the columns in the spectrum's polar matrix. The linear radius (i.e., frequencies) in polar coordinates increases down the rows of the polar matrix. Figures 6.1 and 6.2 display the visible image, thermal image, frequency spectrum, and polar spectrum of a wood wall and brick wall, respectively. Since the discrete Fourier transform used to produce the spectrum assumes the frequency pattern of the image is periodic, a high-frequency

Fig. 6.2 (a) visible image, (b) thermal images, (c) frequency spectrum, and (d) polar spectrum of a brick wall.

drop-off occurs at the edges of the image. These "edge effects" result in intense horizontal and vertical artifacts in the spectrum. Care needs to be taken when generating features from the 2-dimensional frequency domain since these edge effects may interfere with the ability to produce relevant features to classify objects. Fortunately, since these edge effects are consistent for all the thermal images, they will not have a negative impact on sparsity features.

Next, the total energy of the frequencies along the spectral radius is computed for angles from 45 to 224 degrees. This range of angle values ensures that the algorithm captures all possible directions of the frequencies on the object in the scene. A histogram with the angle values along the abscissa and total energy of the frequencies on the ordinate is smoothed using a moving average filter. The values along the ordinate are scaled to obtain frequency energy values ranging from 0 to 1 since we are only interested in how well the edges are defined about the direction of the maximum frequency energy, not the value of the frequency energy. The resulting histogram is plotted as a curve with peaks representing directions of maximum frequency energy. The full width at 80% of the maximum (FW(0.80)M) value on the curve is used to indicate the amount of variation in frequency energy about a given direction. Four features are generated from the resulting histogram defined by the terms: sparsity and direction. The sparsity value provides a measure of how well defined the edge directions are on an object. The value for sparsity is the ratio of the global maximum scaled frequency energy to the FW(0.80)M along a given interval in the histogram. Thus, an object with well defined edges along one given direction will display a curve in the histogram with a global maximum and small FW(0.80)M, resulting in a larger sparsity value compared to an object with edges that vary in direction. To compute the feature values, the intervals from 45 to 134 degrees and from 135 to 224 degrees were created along the abscissa of the histogram to optimally partition the absolute vertical and horizontal components in the spectrum. The sparsity value, along with its direction, is computed for each of the partitioned intervals. A value of zero is provided for both the sparsity and direction if there is no significant frequency energy present in the given interval to compute the FW(0.80)M.

By comparing the directions (in radians) of the maximum scaled frequency energy along each interval, four features are generated: Sparsity about Maximum Frequency Energy (12.03 for wood wall vs. 9.02 for brick wall), Direction of Maximum Frequency Energy (3.14 for wood wall vs. 1.55 for brick wall), Sparsity about Minimum Frequency Energy (0.00 for wood wall vs. 7.80 for brick wall), and Direction of Minimum Frequency Energy (0.00 for wood wall vs. 3.14 for brick wall). Figure 6.3 compares the scaled frequency energy histograms for the wood wall and brick wall, respectively.

As we see in the histogram plot of the wood wall (Fig. 6.3 (a)), the edges are more well defined in the horizontal direction, as expected. Furthermore, the vertical direction presents no significant frequency energy. On the other hand, the results for the brick wall (Fig. 6.3 (b)) imply edge directions that are more well defined in the vertical direction. The brick wall also produces a sparsity value and direction associated with minimum frequency energy. Consequently, these particular results would lead to features that could allow us to distinguish the wood wall from the brick wall.

Fuzzy membership functions could be explored for the sparsity features to translate the vagueness to a degree of membership that produces the "likeliness" of an object being present when given the associated sparsity feature values. It is important to note that the fuzzy logic classifier would be integrated into the Adaptive Bayesian Classification Model to complement the overall classification performance. For instance, the probabilistic (crisp) portion of the model would still recommend a class

264   6 Conclusions and Future Research Directions

assignment along with a posterior probability for an unknown object. However, a fuzzy (non-crisp) portion of the model would fuzzify the sparsity feature values generated from the thermal image of the unknown object to produce an output from the fuzzy set using phrases, such as, *Unlikely* and *Likely*, associated to each object class that could be assigned. For example, for a specific set of sparsity features the fuzzy classifier may output that the unknown object is Likely to be a Wood Wall and Unlikely to be a Brick Wall. The classification model would make a final classification decision based on the recommendations by the crisp and fuzzy classifiers.

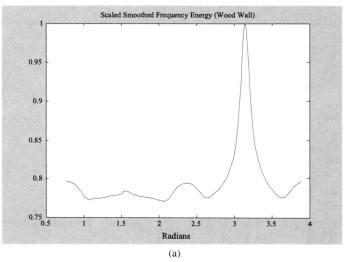

(a)

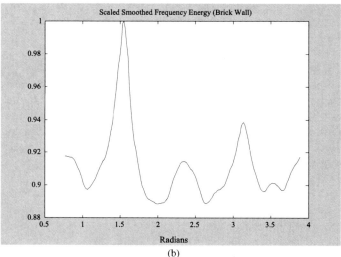

(b)

Fig. 6.3 Scaled frequency energy histograms: (a) wood wall and (b) brick wall.

6.4.3 Bayesian Multi-Sensor Data Fusion

As discussed in Sect. 6.3, the limitations found with any sensor obviously provide the reason why multi-sensor data fusion systems are normally more successful in classification applications than systems with a single sensor. Thus, the interpretations of relevant information received by different types of sensors used in a multi-sensor framework are fused to complement the overall performance of the classification process [7, 8].

Since both ultrasound and infrared are independent of lighting conditions, they are appropriate for use both day and night. Consequently, designing a framework that fuses information from the bot's thermal infrared imaging and ultrasonic sensors for performing intelligent actions, such as decision-making and learning, is an appropriate choice. We envision a Bayesian multi-sensor data fusion architecture involving thermal infrared imaging and sonar sensors as displayed in Fig. 6.4. The first requirement in the multi-sensor data fusion architecture is to ensure the data from the different sensors are registered to common points of reference so that all the sensors are "looking at" the same target. As displayed in the given architecture, the passive thermal infrared imaging and active sonar sensors receive signal data from objects in the surrounding environment. Equivalent to the methodology outlined in this book, the signals received by each sensor are preprocessed to minimize the effects of temporal and spatial signal degradations. The target within the field of view of the sensors is then detected and segmented. After the preprocessing phase, features are generated from the target's signals received by each sensor.

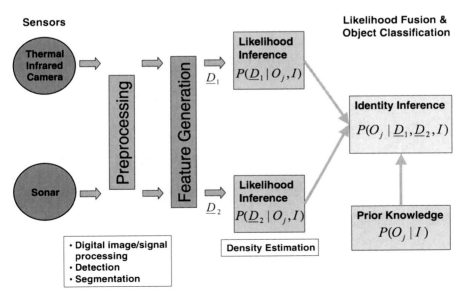

Fig. 6.4 Bayesian multi-sensor data fusion architecture involving thermal infrared and sonar sensors.

The Bayesian multi-sensor data fusion model has the same structure as the Bayesian Classifier discussed in Chap. 4. Thus, the Bayesian multi-sensor data fusion model consists of a likelihood function and prior knowledge to formulate a posterior probability used to classify based on features generated from the unknown target's signals received by each sensor. This logical inference also considers any other relevant background information I. The likelihood function, $P(\underline{D}_n | O_j, I)$, $n = 1,\ldots,M$ and $j = 1,\ldots,J$, provides a measure of the chance that we would have obtained the values in the feature vector $\underline{D}_n$ generated from the unknown target's signal received by sensor n if the object class O_j was given to be present. The prior probability $P(O_j | I)$ provides a measure of our state of knowledge regarding the object class being present before any signal data is collected by the sensors. This prior probability is based on information that we know about the objects in the given environment. If we feel that all the object classes could exist in the bot's local area of operation or have no reason to believe that one object class is more likely to be identified over another, then the "principle of indifference" prevails and we assign equal priors for all the object classes. In Sect. 6.4.4, we will discuss our future research plans to use satellite imagery to assist in developing prior knowledge in a bot's immediate area of operation. Once the likelihood function and the prior probability are established, we use Bayes' theorem to obtain our posterior probability

$$P(O_j | \underline{D}_1,\ldots,\underline{D}_M, I) = \frac{P(\underline{D}_1,\ldots,\underline{D}_M | O_j, I) P(O_j | I)}{\sum_{j=1}^{J} P(\underline{D}_1,\ldots,\underline{D}_M | O_j, I) P(O_j | I)} \quad (6.1)$$

where the unconditional probability $\sum_{j=1}^{J} P(\underline{D}_1,\ldots,\underline{D}_M | O_j, I) P(O_j | I)$ is a normalization parameter (known as the evidence) that ensures $\sum_{j=1}^{J} P(O_j | \underline{D}_1,\ldots,\underline{D}_M, I) = 1$. Since the signals received by the sensors are statistically independent, our likelihood function is computed by $P(\underline{D}_1,\ldots,\underline{D}_M | O_j, I) = \prod_{n=1}^{M} P(\underline{D}_n | O_j, I)$. Thus, with our posterior, we can determine the probability of the target being assigned to object class O_j given the feature vectors generated from the unknown target's signals received each sensor and prior knowledge of the object class existing in the current environment. The Bayesian Multi-sensor Data Fusion Model will assign the target to the object class associated with the largest posterior probability. The Bayesian Multi-sensor Data Fusion Model can be designed in a framework analogous to our adaptive model presented in Chap. 5. Thus, this framework would also include classifica-

6.4 Future Research   267

tion rules that must be satisfied before the bot uses the class assignment to decide on the next required action in the intelligence algorithm.

6.4.4 Prior Knowledge Based on Satellite Imagery

We envision a bot having the ability to use real-time or archival satellite imagery to assist in developing knowledge regarding objects that may exist in an area of operation prior to the bot entering the given area. Hence, the information in the satellite imagery is used to estimate prior probabilities of objects in the bot's immediate area of operation that are used in our Bayesian classification models. We can picture a scenario similar to Fig. 6.5 where a bot, denoted by the blue icon with the given latitude and longitude coordinates, is using satellite imagery to enhance its situational awareness by gaining knowledge of objects that may exist in the next immediate area of operation represented by the region enclosed by the yellow triangle. By partitioning the satellite image into various regions, represented by the enclosed areas with yellow borders and labeled as Paved Road, Yard, and Woods, we are creating surface regions that each consist of a mixture of object classes. For instance, we perhaps know from experience that the region labeled as Woods has a higher chance of containing trees and bushes than fences. The region labeled as Yard could have an equal chance of containing trees, bushes, and fences. On the other hand, the region labeled as a Paved Road could have no chance of containing trees, bushes, or fences. Consequently, we could associate an estimated probability

Fig. 6.5 Autonomous robot estimates prior probabilities of objects in area of operation using satellite imagery to assist in classifying objects within field-of-view of onboard sensors.

for each of these objects existing in each of the respective regions. Thus, as the bot is moving along a specific path, it is conducting a pre-entry analysis of the next area of operation by using satellite imagery to gain prior knowledge of objects that the bot may encounter. The resulting prior probability estimates for each object class from the analyzed region is used in the bot's Bayesian classification model.

Next we need a method to assign a class label (i.e., Paved Road, Yard, and Woods) to the partitioned regions in the satellite image. As we see in Fig. 6.5, each region enclosed by the yellow borders displays a RGB color histogram with distributions that distinguish it from the other regions' histograms. Consequently, we may be able to generate features that uniquely represent the different types of regions that we labeled in the satellite imagery. Suppose the bot captures satellite imagery of the next immediate area of operation in its path, represented by the region enclosed by the yellow triangle, and generates feature vectors from the region's RGB histogram to assign a label to the enclosed region. Estimated prior probabilities are then given for the object classes associated with the respective type of enclosed region. These prior probability estimates are then used as inputs into the bot's Bayesian classification model for computing posterior probabilities of object classes that the bot detects in the next immediate area of operation. If no relevant satellite information is available to predict the region types or there are ties for the type of region, then equal prior probabilities could be assigned for each object class. Additionally, since the partitions are not necessarily crisp in distinguishing region types, we could find a degree of vagueness that may call for the integration of fuzzy logic. Wang [9] describes a fuzzy supervised classification method for classifying land cover in Landsat images involving imprecise boundaries between land cover types. A review of methods used in the classification of remotely sensed data is found in [10].

6.5 Concluding Remarks

We have designed and implemented a physics-based adaptive Bayesian pattern classification model that uses a passive thermal infrared imaging system to automatically characterize non-heat generating objects in unstructured outdoor environments for mobile robots. The resulting model complements an autonomous robot's situational awareness and affords bots with the intelligence to automatically interpret the information in signal data emitted from targets to make decisions without the need for an interpretation by humans. We have demonstrated that our Adaptive Bayesian Classification Model outperforms the traditional KNN and Parzen classifiers.

The framework of our classification model could also be used in other applications requiring the characterization of unknown objects based on features that witness variations due to natural cyclic events. For instance, our model could be integrated into classification applications that use RGB video to generate features from the visible images of objects in outdoor scenes that depend on illumination from the sun. The Adaptive Bayesian Classification Model could also be used during quality control inspections on assembly lines in industry where a thermal pulse is used to stimulate a product's surface and time-varying features generated from the cooling object are used to improve the accuracy of characterizing anomalies in products and monitoring packing standards.

Our work has also laid the foundation for continued research that will: (1) explore the integration of fuzzy logic to assist in classifying targets that emit signal information that imprecisely defines their respective class assignments, (2) design a multi-sensor framework to fuse the interpretations of relevant information received by different types of sensors to complement the overall performance of the classification process, and (3) afford a mobile bot with the ability to use real-time or archival satellite imagery to assist in developing knowledge regarding objects that may exist in an area of operation prior to the bot entering the given area. These interesting and important areas of research are the cornerstone to further advancements in the capabilities of autonomous robotic systems.

References

[1] Holst GC (2000) Common Sense Approach to Thermal Imaging. JCD Pub.; co-published by SPIE Optical Engineering Press, Winter Park, Fla.; Bellingham, Wash.
[2] Maldague XPV (2001) Theory and Practice of Infrared Technology for Nondestructive Testing. Wiley, New York
[3] Hinders M, Gao W et al (FEB 2007) Sonar Sensor Interpretation and Infrared Image Fusion for Mobile Robotics. In: Kolski S (ed) Mobile Robots: Perception & Navigation. Pro Literatur Verlag, Germany / ARS, Austria
[4] Goubet E, Katz J et al (May 2006) Pedestrian tracking using thermal infrared imaging. Proceedings of SPIE, Infrared Technology and Applications XXXII:62062C-1 - 62062C-12
[5] Ayyub BM, Klir GJ (2006) Uncertainty Modeling and Analysis in Engineering and the Sciences. Chapman & Hall/CRC, Boca Raton, FL
[6] MathWorks (2007) Fuzzy Logic Toolbox 2: For use with MATLAB: User's Guide. 2nd edn. MathWorks, Inc., Natick, MA
[7] Hall DL, McMullen SAH (2004) Mathematical Techniques in Multi-Sensor Data Fusion. 2nd edn. Artech House, Boston
[8] Hall DL, Llinas J (2001) Handbook of Multisensor Data Fusion. CRC Press, Boca Raton, FL
[9] Wang F (1990) Fuzzy Supervised Classification of Remote Sensing Images. IEEE Transactions on Geoscience and Remote Sensing 28(2):194–201
[10] Kokhan S (2007) Classification of Remotely Sensed Data. Geographic Uncertainty in Environmental Security:239–247

Index

A

Absolute posterior difference threshold, 227
AC coupling, 33, 54, 62
Acoustics, *See* Ultrasound
Adaptive Bayesian classification model, 161, 223, 227, 260
Adaptive Bayesian classifier, 193, 197
Ambient temperature, 19, 54
Ambient temperature rate of change, 54
Angular second moment, *See* Energy
Attributes, *See* Features
Automatic gain control (AGC), 34
Automatic target recognition (ATR), 16
Autonomous robotic system, 8

B

Background, 27, 53
Background irradiance, 53
Bayes' formula, *See* Bayes' theorem
Bayes' theorem, 106
Bayesian classifier, 105, 193
Binomial law, 107
Blackbody, 17, 27
Blind data set, 42, 122
Bootstrap, 122

C

Chopper wheel, 37, 60
Classification model, 7, 13, 25, 95, 116
    observational model, 14
    theoretical model, 14
Classification rules, 226

Classify, 13
Combining classifiers, 223
Committee mode frequency threshold, 226
Committee of experts, 162, 223
Compact objects, 41, 47, 95, 161
Component (or scalar projection), 164
Conditional probability, 105
Conductivity, 19
Confusion matrix, 121
Contrast, 60, 71, 76
Control IR Manager, 27
Correlation, 76
Cross-validation, 113, 122
Curse of dimensionality, 116
Curvature algorithm, 88, 256

D

DARPA Grand Challenge, 11
Data dredging, 96
Decision boundary, 97
Density estimation
    nonparametric, 97, 98, 105
    parametric, 97
Detection, 11
Discriminant functions, 97
Distance function, 196
Dynamical window, 150, 162, 223, 257

E

Edge effects, 262
Emissivity, 18, 55, 58, 59, 63, 257
Emittance, 55
Energy, 77, 78, 80, 81, 82, 86, 87, 124, 134

Entropy, 72, 73, 74, 78, 80, 81, 82, 86, 88, 124, 134
Error estimation, 113, 122
 bootstrap, 122
 cross-validation, 113, 122
 holdout, 122
 leave-one-out, 122, 123
 resubstitution, 122
 rotation, 122
Error (or misclassification) rate, 113, 120, 219
Evidence, 106, 266
Exhaustive search, 84, 115, 120
Extended objects, 41, 66, 95, 161

F

FastICA, 162
Feature extraction, 116
Feature selection, 119
Feature vector (or pattern), 96
Features, 7, 14, 47
 geometric, 7, 48
 thermal-physical, 48
First principal eigenvector, 161
First-order statistical features, 70
Focal plane array (FPA), 13, 27
Foreground, 27, 53, 144
Fourier-Mellin descriptors, 48
Fourier transform, 261
Frequency spectrum, 260, 261
Fuzzy logic, 222, 260, 263, 268

G

Geometric features, 7, 48
Graybody emitter, 55
Gray-level co-occurrence matrix, 74
Gray-scale (or gray-level) values, 18, 29, 33, 34, 35, 36, 51, 53, 54, 59, 60, 61, 62, 66, 69, 70, 71, 72, 73, 74, 75, 76, 77, 87, 89, 213

H

Halo effect, 37, 60, 88
Heat transfer mechanisms, 19
 conductive, 19
 convective, 19
 radiative, 19
Heating flux, 19
High pass filter, 39
Holdout method, 122
Homogeneity, 77, 80

Hu's seven moments, 48
Hyperconoidal cluster, 105, 114, 116, 119, 121, 161, 162, 163, 164, 193, 194, 198, 210, 216, 219, 224, 251, 253, 256, 257

I

Inertia, *See* Contrast
Infrared range sensor, 11
Infrared thermography, 15
Inter/intra class distance, 103, 120
Irradiance, 53, 59

J

Jacknife method,
 See Leave-one-out method
Joint probability, 105

K

Kirchhoff's law, 55
K-Nearest-Neighbor (KNN) classifier, 110, 197
KNN density estimation, 109, 111, 194, 196

L

Laser detection and ranging (LADAR), 11
Leave-one-out method, 122, 128
Likelihood function, 106, 193, 266

M

Machine vision, 8, 50
Macro features, 69, 74
Maximum frequency energy, 263
Meteorological features, 54, 85
Micro features, 55
Misclassification matrix,
 See Confusion matrix
Misclassification rate, *See* Error rate
Mode class, 226
Multi-mode heat transfer, 18
Multi-sensor data fusion, 265

N

Nearest neighbor rule, 111
Neural networks, 96
No Free Lunch Theorem, 96, 223
Nondestructive evaluation (NDE), 16
Non-heat generating objects, 2, 27, 95

Nonparametric density estimation, 97, 98, 105
  decision boundary, 97
  probabilistic, 97, 98
Nonparametric techniques, 97
Normal distance, 161, 257
Normalization correction, 30
  non-uniformity correction, 30

O

Object scene radiance, 70
Object surface radiance, 51

P

Parametric density estimation, 97
Parzen classifier, 111, 197
Pattern, *See* Feature vector
Pattern classification (or recognition), 14, 16, 48, 96
Pattern classification model, *See* Classification model
Pattern recognition, *See* Pattern classification
Peaking phenomenon, 116, 124, 134
Perfect emitter, *See* Blackbody
Performance criterion, 120
Pixel distance, 74
Pixel substitution, 30
Planck's blackbody radiation law, 27
Planck's law, 17
Polarity, 33
Polar spectrum, 261
Posterior probability, 106, 196, 266
Posterior threshold, 226
Principal axis, 118
Principal component, 119
Principal component analysis (PCA), 117, 162
Principle of indifference, 106, 114, 266
Prior probability, 106, 110, 194, 266
Probability density function, 107
PRTools4, 162

R

Radiance, 48
Radiosity, 53
Raytheon ControlIR 2000B, 26
Redundancy reduction, 99
Reference emitter radiance, 65
Required class votes threshold, 226
Resubstitution method, 122

RGB cameras, 12
Rotation method, 122

S

Samsung Tablet PC, 27
Satellite imagery, 266
Scalar projection, *See* Component
Second-order statistical features, 74
Segment, 11
Sensors, 11
  active sensors, 11
  passive sensors, 11
Shannon's entropy, *See* Entropy
Signal degradations, 28
  dead pixels, 28
  spatial, 28
  temporal, 28
Situational awareness, 1
Smoothing parameters, 113
Smoothness, 71
Sonar, *See* Ultrasound
Sparsity, 260
Specific heat, 19, 54, 258
Statistical pattern classification, 96, 97, 98
Stephan-Boltzmann coefficient, 19
Supervised classification, 97
Synthetic aperture radar (SAR), 11

T

Template matching, 96, 97
Terahertz-pulsed imaging, 12
Terrain classification, 14
Test data set, 41, 115, 121
Texture, 51
Thermal (long-wave) infrared, 17
Thermal (long-wave) infrared detector, 17
Thermal crossover, 18, 101, 222, 257
Thermal infrared imaging, 16
  active thermal infrared imaging, 16
  passive thermal infrared imaging, 16
Thermal-physical features, 48
  macro, 69, 74
  meteorological, 54, 85
  micro, 55
Thermographic nondestructive testing (TNDT or NDT), *See* Nondestructive evaluation (NDE)
Thermography, 15
Third moment, 72
Thresholds
  absolute posterior difference, 227
  committee mode frequency, 226

posterior, 226
required class votes, 226
ties, 226
Ties, 114
Ties threshold, 226
Training data set, 41, 115, 121

U

Ugly Duckling Theorem, 48
Ultrasound, 5, 11, 12, 13, 15, 222, 259, 265

Unconditional probability, 196, 266
Uniformity, 72, 87
Unsupervised classification, 97

V

Validation data,
    See Test data set

Z

Z5 standardization method, 98